# Creative Thinking in University Physics Education

Online at: https://doi.org/10.1088/978-0-7503-4028-1

# IOP Series in Physics Education

The IOP Series in Physics Education aims to provide comprehensive, authoritative and innovative coverage for those that teach physics and related subjects at universities and other higher and further education institutions, and for those involved in physics education research.

**Series Editor**
**Professor Peter Main**
*King's College London, UK*

**About the Editor**
Peter Main obtained his PhD from the University of Manchester and, after post-docs in Manchester and Helsinki, he joined the University of Nottingham as a Lecturer in Physics in 1979. Following promotions to Reader and Professor, he eventually became Head of the School of Physics and Astronomy. His principal research interests were in quantum fluids and quantum transport in semiconductor and metallic heterostructures. He was also involved in many teaching innovations.

In 2002 he left Nottingham to join the Institute of Physics as Director of Education and Science. In this post he had overall responsibility for the Institute's work in education at all age levels, research, and diversity. Among many projects, he worked closely with Ofqual and awarding bodies on curriculum matters and with government to increase the number of physics teachers. He also initiated several projects improving the diversity of participation in physics.

In 2015 he joined King's College to become Head of Physics; he retains his interest in many projects in physics education and diversity.

**About the Series**
The IOP Series in Physics Education aims to provide comprehensive, authoritative, and innovative coverage for those that teach physics and related subjects at universities and other higher and further education institutions, and for those involved in physics education research.

The series supports evidence-informed professional practice and will cover topics including the following: assessment methods; feedback; conceptual understanding; problem solving; teaching methods; education technology; pedagogical theory; curriculum design; student engagement; misconceptions; employability; and social aspects of education.

Authors are encouraged to take advantage of electronic publication through the use of colour, animations, video, data files, and interactive elements, all of which offer particular benefits in communicating pedagogy.

**Do you have an idea for a book you'd like to explore?**
We are currently commissioning for the series; if you are interested in writing or editing a book please contact Caroline Mitchell at caroline.mitchell@ioppublishing.org.

A full list of titles published in this series can be found here: https://iopscience.iop.org/bookListInfo/iop-series-in-physics-education

# Creative Thinking in University Physics Education

**Douglas P Newton, Sam Nolan and Simon Rees**
*School of Education, Durham University, Durham, UK*

**IOP** Publishing, Bristol, UK

© IOP Publishing Ltd 2022

All rights reserved. No part of this publication may be reproduced, stored in a retrieval system or transmitted in any form or by any means, electronic, mechanical, photocopying, recording or otherwise, without the prior permission of the publisher, or as expressly permitted by law or under terms agreed with the appropriate rights organization. Multiple copying is permitted in accordance with the terms of licences issued by the Copyright Licensing Agency, the Copyright Clearance Centre and other reproduction rights organizations.

Permission to make use of IOP Publishing content other than as set out above may be sought at permissions@ioppublishing.org.

Douglas P Newton, Sam Nolan and Simon Rees have asserted their right to be identified as the authors of this work in accordance with sections 77 and 78 of the Copyright, Designs and Patents Act 1988.

ISBN    978-0-7503-4028-1 (ebook)
ISBN    978-0-7503-4026-7 (print)
ISBN    978-0-7503-4029-8 (myPrint)
ISBN    978-0-7503-4027-4 (mobi)

DOI    10.1088/978-0-7503-4028-1

Version: 20221201

IOP ebooks

British Library Cataloguing-in-Publication Data: A catalogue record for this book is available from the British Library.

Published by IOP Publishing, wholly owned by The Institute of Physics, London

IOP Publishing, No.2 The Distillery, Glassfields, Avon Street, Bristol, BS2 0GR, UK

US Office: IOP Publishing, Inc., 190 North Independence Mall West, Suite 601, Philadelphia, PA 19106, USA

# Contents

| | | |
|---|---|---|
| **Preface** | | x |
| **Author biographies** | | xii |

| **1** | **Creative thinking in physics** | **1-1** |
|---|---|---|
| 1.1 | The relevance of creative thinking | 1-1 |
| 1.2 | Physics: cold comfort farm or possibility place? | 1-2 |
| 1.3 | Creative students? | 1-4 |
| 1.4 | Creative thinking doesn't come with a guarantee | 1-6 |
| 1.5 | Variety, the spice of physics teaching | 1-7 |
| 1.6 | Does it matter? | 1-9 |
| 1.7 | Something to reflect on | 1-11 |
| | References | 1-11 |

| **2** | **The creative learner in physics** | **2-1** |
|---|---|---|
| 2.1 | Learning physics and learning what counts in physics | 2-1 |
| 2.2 | Noticing and making sense of problems | 2-2 |
| 2.3 | Constructing understandings to enable explanation | 2-5 |
| 2.4 | Testing a tentative explanation or idea | 2-6 |
| 2.5 | Application | 2-8 |
| 2.6 | Creative thinking is not a mechanical process | 2-9 |
| 2.7 | Something to reflect on | 2-10 |
| | References | 2-10 |

| **3** | **Creative thinking in practice: problems** | **3-1** |
|---|---|---|
| 3.1 | Fertile problems | 3-1 |
| 3.2 | Curiosity and questions | 3-1 |
| 3.3 | Noticing, finding, and posing problems | 3-2 |
| 3.4 | The problem of eliciting students' questions | 3-4 |
| 3.5 | Fostering students' thinking about problems | 3-6 |
| | 3.5.1 A problem analysis heuristic | 3-7 |
| | 3.5.2 A qualitative account heuristic | 3-7 |
| | 3.5.3 Working backwards | 3-7 |
| | 3.5.4 Using an analogy | 3-7 |
| 3.6 | The tutor's contributions | 3-8 |
| 3.7 | There is no end to questions | 3-9 |

| | | |
|---|---|---|
| 3.8 | Something to reflect on | 3-9 |
| | References | 3-10 |

## 4  Creative thinking in practice: ideas — 4-1

| | | |
|---|---|---|
| 4.1 | Introduction | 4-1 |
| 4.2 | Astronomer Copernicus | 4-1 |
| | 4.2.1  Student activity—structures in our mind | 4-2 |
| 4.3 | Divergent thinking | 4-3 |
| | 4.3.1  Student activity—divergent thinking | 4-4 |
| 4.4 | Convergent thinking | 4-4 |
| | 4.4.1  Student activity—*The Martian* | 4-5 |
| 4.5 | Associative thinking | 4-6 |
| | 4.5.1  Student activity—new connections | 4-6 |
| 4.6 | Effective ideas generation | 4-6 |
| 4.7 | Lateral thinking | 4-7 |
| | 4.7.1  AlphaGo | 4-8 |
| | 4.7.2  Problem based learning | 4-9 |
| 4.8 | Sticky creativity | 4-9 |
| | 4.8.1  The curling conundrum | 4-11 |
| 4.9 | Conclusion | 4-11 |
| 4.10 | Something to reflect on | 4-12 |
| | References | 4-12 |

## 5  Creative thinking in practice: experiments — 5-1

| | | |
|---|---|---|
| 5.1 | Introduction | 5-1 |
| 5.2 | The affective domain | 5-1 |
| 5.3 | Gender equity | 5-2 |
| 5.4 | Experimental demonstrations | 5-3 |
| | 5.4.1  Faraday and Tyndall at the Royal Institution | 5-4 |
| 5.5 | Objects as analogies and metaphors | 5-5 |
| | 5.5.1  Student activity—everyday objects | 5-6 |
| | 5.5.2  Tutor activity—challenging concepts | 5-7 |
| 5.6 | Thought experiments | 5-7 |
| 5.7 | Inquiry based learning | 5-8 |
| | 5.7.1  Case study—experimental creativity | 5-9 |
| | 5.7.2  Project based learning | 5-10 |
| | 5.7.3  Tutor activity—project based learning | 5-11 |

|       |       | 5.7.4 Technicians | 5-11 |
|---|---|---|---|
| 5.8 | Something to reflect on | | 5-12 |
|     | References | | 5-12 |

## 6 Creative thinking in practice: applications — 6-1

| 6.1 | Introduction | 6-1 |
|---|---|---|
| 6.2 | Frameworks for creativity in learning | 6-2 |
|     | 6.2.1 A framework for the creative process | 6-2 |
| 6.3 | Designing a creative learning activity | 6-3 |
| 6.4 | Case studies | 6-4 |
|     | 6.4.1 Conceiving a guesstimation based curriculum (University of Glasgow) | 6-4 |
|     | 6.4.2 Team projects at Durham University | 6-6 |
| 6.5 | Discussion | 6-8 |
| 6.6 | Conclusions | 6-10 |
| 6.7 | Something to reflect on | 6-12 |
|     | References | 6-12 |

## 7 Recognising creative thinking in physics — 7-1

| 7.1 | Uncertainty and assessing thinking competences | 7-1 |
|---|---|---|
| 7.2 | Assessing the product of creative thought | 7-2 |
|     | 7.2.1 Assessment by consensus | 7-2 |
|     | 7.2.2 Simple rating | 7-3 |
|     | 7.2.3 Putting numbers to it | 7-3 |
| 7.3 | Assessing the process of creative thought | 7-5 |
|     | 7.3.1 Just ask | 7-5 |
|     | 7.3.2 In writing | 7-6 |
|     | 7.3.3 Observe | 7-6 |
| 7.4 | Working in groups on practical and other tasks | 7-6 |
|     | 7.4.1 Collaborative competence | 7-7 |
|     | 7.4.2 Group creative competence | 7-7 |
| 7.5 | Risk taking and some caveats | 7-8 |
| 7.6 | Providing feedback | 7-10 |
| 7.7 | Recognition and measurement | 7-10 |
| 7.8 | Something to reflect on | 7-11 |
|     | References | 7-11 |

# 8  The creative tutor — 8-1

8.1  The value of creative teaching — 8-1
8.2  Ten questions and answers — 8-1
    8.2.1  What is creative teaching in the context of higher education (HE)? — 8-2
    8.2.2  What are the benefits of creative teaching? — 8-2
    8.2.3  Is there a downside to teaching creatively? — 8-3
    8.2.4  What would you say are the attributes of a creative teacher in HE? — 8-3
    8.2.5  Are these attributes something you are born with, or can they be acquired or developed? — 8-4
    8.2.6  What are the impediments to creative teaching? — 8-4
    8.2.7  Students sometimes see themselves as buying a product rather an education. Do you think this adversely affects a desire to teach creatively? If so, in what way(s)? — 8-5
    8.2.8  What advice would you give to a new university lecturer about becoming a more creative teacher? — 8-5
    8.2.9  Would you give different advice to someone who is a mid-career lecturer? — 8-6
    8.2.10  Do you see higher managerial colleagues in a university as having a role in fostering creative teaching? If so, what would that role be? — 8-6
8.3  Why teach creatively? — 8-6
    8.3.1  Some benefits for the student — 8-6
    8.3.2  Some benefits for the tutor — 8-6
8.4  Creative teaching to support students' learning — 8-7
8.5  Creative uses of technology — 8-8
    8.5.1  Solving teaching and learning problems — 8-8
    8.5.2  Technology taking some of the strain — 8-9
8.6  The place of critical/evaluative thinking — 8-9
8.7  Change and challenges — 8-9
8.8  Some things to reflect on — 8-10
    References — 8-11

# 9  Creative approaches to teaching physics in the twenty-first century — 9-1

9.1  Laboratory learning — 9-1
    9.1.1  Interactive simulated experiments — 9-2
    9.1.2  At home laboratories — 9-4

| | | |
|---|---|---|
| 9.2 | Simulation based learning | 9-6 |
| 9.3 | The use of virtual and augmented reality in physics teaching | 9-7 |
| 9.4 | Enhancing peer learning in lectures with technology | 9-8 |
| 9.5 | Judging support tools | 9-10 |
| 9.6 | The future | 9-10 |
| 9.7 | Something to reflect on | 9-11 |
| | References | 9-11 |

## 10  Creating change     10-1

| | | |
|---|---|---|
| 10.1 | Taking the wider view | 10-1 |
| 10.2 | Some roles | 10-1 |
| 10.3 | Some hurdles | 10-3 |
| | 10.3.1 Inertia and fragmentation | 10-3 |
| | 10.3.2 Notions of creativity | 10-4 |
| | 10.3.3 Notions of the source of creative abilities | 10-4 |
| | 10.3.4 Students' notions of creative physics | 10-5 |
| 10.4 | Health, safety, and risk assessment | 10-5 |
| 10.5 | Physics as a dynamic discipline | 10-6 |
| 10.6 | Creative physics and the cultivated imagination | 10-6 |
| | References | 10-7 |

# Preface

Popularly, physics is not seen as a creative activity, but as a subject that searches for obscure facts, notes laws of nature, and produces and manipulates difficult formulae. For many, what they see as a tedious pursuit produces a body of knowledge to which new information is simply added, with little room for personal, constructive contributions. With an image like this, it is hard to compete with the many attractions on offer elsewhere. We may even reinforce the image by hiding creative thinking behind the term 'problem solving', and then substituting closely defined exercises for the problems that really do need imagination to solve them. We know it to be a false image, and it's one we can do something about. Competence in creative thinking is an asset and is likely to be at a premium in the future. We should foster it in our students, but we are not likely to do so if we don't give it overt attention or have students exercise it. Given opportunities, students can find it attractive, rewarding, and motivating.

Sometimes we also need to change our own mindset. It may be true that a few, lucky students have a genetic makeup which favours their imagination, but for most of them—and us—interests, skills, and thinking competences come from a blend of nature and nurture, what we are born with and the experiences we have. We cannot dismiss someone as simply 'not having what it takes to be creative' without first shaping their experience to allow it to develop. We also need to have a view of what we mean by creative thinking in physics. There can be a tendency to think that it is only present when its product is successful, or somehow 'correct'. Creative thinking in physics needs to produce something plausible, but plausibility is not a guarantee of ultimate truth. Plausible ideas may be found wanting but that is not to say that they are unimaginative ideas which had no potential when they were constructed. The history of physics is full of such ideas, and when we were planning our book we were urged by reviewers to draw on the history of scientific ideas to illustrate and make such points clear. On several occasions, particularly in the early chapters, we found this very useful, and suspect that readers will be able to supplement these with their own experiences. Sharing these experiences with your students can engage them and help them see physics as a dynamic activity, rather than as a compendium of unchanging knowledge that has to be acquired.

Clearly, we see value in fostering students' competence in creative thinking. The world is changing rapidly, and much of that change has its origins in the evolving body of knowledge we call science, and the application of that knowledge to solve practical problems. This demands creative thinking. But we do not believe that other kinds of thinking are unimportant or can be neglected. Recall, understanding, deduction, analysis, and evaluation are all important in themselves. Creative thinking needs them too, if it is to be productive. We have seen students' interest and engagement in even small opportunities for them to be imaginative and to produce something new, at least to them. Not only do such opportunities keep alive a way of thinking which will be of benefit to them (and us), but they find more satisfaction in what they do. Both are worthwhile

Teaching practices obey Newton's first law. They often have an enormous inertia and resist change. Changes in perspective may be needed, including at the highest level if teaching is to reflect the nature of a discipline, and add to a students' assets. And those who strive to make students' learning more relevant and creative need to be heard, valued, supported, and encouraged. They also need opportunities to be creative in their teaching themselves so that they bring a fresh and more dynamic experience to their students.

A final reminder is that creative thinking can produce a wide range of ideas, and some may be not be anticipated. **Do not neglect the need for risk assessments of practical activities.**

Douglas P Newton PhD DSc<br>
Sam Nolan PhD<br>
Simon Rees PhD

# Author biographies

### Douglas Newton

**Professor Douglas Newton** PhD DSc SFHEA has a background in physics education, and specialises in research on science education, engagement, emotions and learning, and creative thinking, particularly in the sciences. He has written or co-authored many papers and some 50 books, and has several active research projects on creativity in the sciences. He was invited to contribute a history of his work to the 2019 volume of the *International Journal for Talent Development and Creativity* (**6** 211–9), titled 'Profiles of creativity: an autobiography of Douglas P Newton'.

### Sam Nolan

**Professor Sam Nolan** PhD (Phys), MSci, NTF, PFHEA is Deputy Director at the Durham University Centre for Academic Development where he works on strategic projects on teaching enhancement. Sam has a background as a physics researcher and teacher, with scholarship interests including widening participation into higher education, virtualisation of laboratory learning, augmented reality, and the use of active learning pedagogies to enhance student engagement.

### Simon Rees

**Professor Simon Rees** PhD (Chem), PhD (Educ) has a university-wide responsibility for the development of postgraduate students and academic staff. He has written on widening participation, educational research, and he is the main co-author of a well-received book on being creative in chemistry.

IOP Publishing

# Creative Thinking in University Physics Education

Douglas P Newton, Sam Nolan and Simon Rees

# Chapter 1

## Creative thinking in physics

### 1.1 The relevance of creative thinking

Digital technologies are set to change lives. Several studies predict that digitally enabled devices will displace people from the work they currently do. The prediction varies from study to study with the optimists suggesting that different work will be generated to replace that which is lost, while the pessimists foresee mass unemployment (Frey and Osborne 2013, Bakshi *et al* 2015, Arntz *et al* 2016, Ramge 2019, Clark 2020). Occupations that are more likely to survive, however, are those calling for creative thought and action (some of which may be augmented by artificial intelligence), and so creative competence is likely to be at a premium and will be a marketable commodity for graduates because of the intellectual property it may produce. A worldwide consequence is an interest in fostering creative competence in education in all domains and at all levels.

Subjects such as physics are popularly seen as being outside creativity—there are the 'creative arts' but no-one talks of the 'creative sciences'. This is partly due to the way they are presented to the world. The shop front is neatly set out with highly polished theories, evidence, explanations, and applications, while the back room has the real and messy story of people's creative thoughts, imaginative solutions, and clever ideas (and those that turn out not to be solutions or quite as clever as they seemed at the time). But the shop window that presents physics as a static, finished subject tends to be what students see, and undergraduates' perceptions of various aspects of physics may hardly change over their time at university (Bates *et al* 2011). This risks repelling those who want to invest something of themselves in a work in progress, a dynamic subject capable of change (deWitt *et al* 2019). However, it can also misrepresent the subject, even to those who study physics but never see its creative side, and it ignores a world in which creative competence will be increasingly important and a valuable asset (Bao and Koenig 2019).

Here we emphasise that we do not say that what is on display in the shop window should not be there, but what we do say is that we should not forget to talk about the back room. An honest presentation of the subject can give students a chance to see and appreciate something of the true and creative nature of the scientific process as it happens in physics. Yes, students should continue to exercise their deductive thinking and their critical or evaluative thinking, but they should not neglect creative thinking. Exercising creative thinking and showing some competence in it adds to the value of the package that learning physics offers and which some may carry forward on their way to becoming professional physicists.

## 1.2 Physics: cold comfort farm or possibility place?

There is a popular belief that subjects such as physics simply reveal by cold observation and deduction what is already there. Some postgraduates we worked with recently were surprised that our topic was creativity, then they were drawn in, even excited by the possibility that what they did could be a part of the creative endeavour of science. They also said that, in their undergraduate years, they rarely had the opportunity or freedom to be scientifically creative. Instead, they had laboured over memorising facts, laws, theories, formulae, and finding the 'right' answer to pre-set problems for which the answer is already known. Sadly they had not been given the opportunity to experience this highly motivating, engaging, and central activity in the scientific process of knowledge creation. This is not just our experience. There is widespread evidence that there is little deliberate fostering of creative abilities in higher education at either undergraduate or postgraduate levels in many disciplines (e.g. Jackson and Shaw 2006, MacLaren 2012, Rogaten and Moneta 2016, Egan *et al* 2017, Jahnke *et al* 2017, Marquis *et al* 2017, Brodin 2018). But, of course, physics really is a creative endeavour. The vignette from the history of our notion of the atom clearly shows this. It is a story of using the imagination to make sense of what cannot be seen, and it can offer useful messages for undergraduates.

### Understanding the atom

A puzzle since the days of antiquity has been the nature of matter. Is it essentially continuous and divisible into smaller and smaller pieces of itself? Or is there a limit when a substance is finally divided into its smallest piece beyond which it is impossible to go? In 1803 this long-standing question attracted the attention of John Dalton. He noted that compounds always comprised simpler substances in the same proportion by mass, and he found he could explain this if he imagined them to be made up of small, indestructible, indivisible particles or atoms. This billiard ball model of the atom was plausible and elegantly explained other features of substances.

However, in 1898 J J Thomson was studying the glow emitted by the cathode in discharge tubes. He found that its nature did not depend on the material of the cathode, it was attracted to a positively charged plate, and that it behaved like a particle which was only about a two-thousandth of the mass of the lightest atom, the hydrogen atom. The problem was that this did not fit Dalton's model of the indivisible atom.

Thomson's solution was to imagine that, instead, the atom comprised a positively charged body with negative 'corpuscles' (electrons) embedded in it, what became known as the plum pudding model. This was about the time when radioactive substances and alpha particles were discovered.

A few years later (1911) Ernest Rutherford had these positively charged alpha particles fired at gold foil. He expected that they would pass through largely unaffected by the gold plum pudding atoms, but, to his surprise, some rebounded. (He described it as being like firing a 16 inch (40 cm) shell at tissue paper and observing it bounce back.) He explained this unexpected effect by imagining that the positively charged part of the atom was concentrated in a core or nucleus and the negatively charged electrons were dispersed to orbit around it: his solar system model of the atom.

This model was soon refined by Niels Bohr (1913) who confined the electron to specific orbits in order to explain the specific colours of light radiated by the hydrogen atom. Franck and Hertz (1914) found experimental evidence consistent with this model, but, unfortunately, it was found to be less successful in explaining the colours of light emitted by other atoms, and it was subsumed by a quantum mechanical model of the atom which removed the notion of definite orbits.

However, this still left the nature of the nucleus of the atom as a puzzle. By this time it was known to have protons and neutrons. Were they simply mixed together in another plum pudding? If so, why are some nuclei very stable and others are not? Around 1949 Maria Goeppert Mayer and others imagined the nucleus to be organised in neutron and proton 'shells' which, if complete, give the nucleus significant stability. This model explained the 'islands' of stability in the periodic table.

And so it goes on as the desire to understand and explain the physical world drives the imagination to produce mental structures which explain it.

The vignette shows that attempts to understand the nature of matter are essentially creative—they are *our* notions, *our* inventions, *our* constructions of the mind, but not just any crazy constructions, they have to be plausible, and it is good if they are also somehow economical, elegant, clever or satisfying constructions, although they may later become less satisfying as weaknesses reveal themselves and prompt the creation of other models. The American educator John Dewey saw teaching as the process of supporting the ability to construct mental structures. Creating such understandings is one of the central aims of physics. Testing them is another, even when the result is expected to be consistent with predictions, as was the case in the alpha particle investigation instigated by Rutherford. Practical investigations may also serve other ends and call for very creative thinking, as with Robert Millikan's method for determining the size of the charge on Thomson's electron, or the Franck and Hertz evidence in support of the Bohr atom. Then there is the use made of an understanding to explain events or facilitate actions in other elements of physics, as in the use made of the knowledge of the electron in electronics. It may also be in the development of a new tool, such Le Châtelier's use of the thermocouple to measure high temperatures. Or it may be in the formulation of a concept, as with the notion of *vis viva*, replaced by the more useful concept we call kinetic energy (although it still survives in the orbital-energy-invariance law). Another example is the Feynman diagram (sometimes referred to as a Stueckelberg diagram after the pioneer of the idea) used in quantum mechanics to show concisely and meaningfully what happens when elementary particles collide (Wilczek 2016). Before any of this, however, there needs to be an awareness or noticing of a problem, a puzzle, or a paradox and an inclination to try to solve it. This is not always easy. Getting a handle on a problem can involve seeing it from a different perspective and thinking in a 'What if?' world. Paul Dirac speculated about the existence of a magnetic monopole, a free particle with a single magnetic north or south 'charge'. None has been found, so is this an opportunity for a would-be physicist, a problem which needs an imaginative interpretation of what a magnetic monopole means?

Such thinking generally involves:
- Producing something more or less **new**, **novel**, or **original**;
- That something should be **appropriate**, **fit for purpose**, and **plausible**; and,
- Preferably, it should be **parsimonious**, **elegant**, **impressive**, **clever**, or otherwise **satisfying**.

There are many definitions of creative thinking, but Acar *et al* (2017) found that they can generally be reduced to these three elements. The first two of these are essential and have been loosely expressed as Creativity = Originality × Appropriateness: if either term is zero, then the whole is zero (Simonton 2012). An economical, elegant, or otherwise satisfying product is desirable, it adds attraction but is not essential (Trefil 2002). What creative thinking excludes, however, is routine, reproductive, or formulaic thinking, and implausible novelty. But, as the vignette exemplifies, the creative process is rarely a once and for all event.

## 1.3 Creative students?

Students, however, sometimes say disconsolately that they are not creative. At times this comes from a notion of creativity that is tied to the arts, but this reflects the way words tend to be used. The word *creativity* tends to be associated with activity in the arts while in the sciences, mathematics, and technology, the practice is to refer to it as *problem solving*. In the vignette, the problem was to understand the fundamental nature of matter. The solution was to construct a model of it, subject to what was known. The act of problem solving is essentially a creative one, calling for a kind of thinking that brings together ideas to form something new. (Even in the arts, when the activity is subject to tight constraints, the act may be called problem solving, as when a composer creates a piece of music for a particular event.)

There can also be an assumption that competence in creative thinking is an innate trait that some lucky people have and can use anywhere. In reality, we may all be able to think more or less creatively, but an inclination to use what we have may be stifled or starved by the way we are taught and by a lack of opportunity, practice, and expectation. In addition, creative competence can be 'domain specific': we may be more creative in physics than in chemistry, and more creative in both than in, say, history or running a business (Lebedeva *et al* 2019). Some of this selective creative tendency may, of course, reflect experience and interest but, given students who willingly choose to study physics, we might expect them to be more disposed to act creatively in it, given the chance, than they would in some other discipline.

Sometimes students believe that creative thinking implies that they are expected to produce something worthy of a Nobel Prize, and some readers may ask, 'What can students produce that is novel, new, or original in a field where experienced physicists have laboured for years without success?' First, a professional physicist is expected to produce something generally seen as *new to the world* and *fit for purpose*. Students, however, are not professional physicists and this has to be recognised. They may not produce something entirely new to the world, but they may create something that is at the least *new to them* and *fit for purpose*, commensurate with their experience and knowledge, and rewarding or *satisfying* for them as a personal kind of creativity (Kaufman and Beghetto 2009). Most students have some potential to be creative, but they may not know it, show it, know it is welcome, or else have developed habits of thought which ignore it. For instance, there is a personal creativity that 'is inherent in any act of learning'. This occurs when students construct new insights or understandings that can make their learning meaningful. These may not be new to the world but are new to the student. Noticing something puzzling in the physical world, or given a problem relating to the course work, students may explore it and construct explanations as novice physicists. In due course, they may address problems like those of the professional physicist who has accumulated knowledge and experience in some aspect of a subject and whose work is open to public scrutiny. Only rarely can we expect them to go beyond that and engage in a major creative activity that most agree changes perspectives on a topic (as with the kinetic theory) and, in the process, join the ranks of the eminent physicist. While, in theory, an undergraduate student may perform at any of these levels, in practice, it is more likely to involve personal and novice creativity. (There are always exceptions. Physicists, like mathematicians, tend to produce their best work when relatively young, unhindered by prevailing assumptions, so it is not out of the question that a student may produce something new to the world. Lawrence Bragg's study of the arrangement of atoms in molecules using x-ray diffraction patterns is an example.) The postgraduate, however, is usually expected to show some professional creativity. Figure 1.1 serves to indicate the kinds of creativity and their incidence that we might expect of a student. These kinds of creative thinking are neither mutually exclusive nor as sharply defined as figure 1.1 may suggest (e.g. Beghetto and Kaufman 2009, 2015). Creative thinking comes in all shapes and sizes, and it is for tutors to know their students and provide suitable opportunities for them to exercise and develop all their thinking, creative and otherwise. The point is that students may operate at various levels of creative thinking and, while we may anticipate that much will be personal and novice-like, it can extend into the professional and be novel to the world.

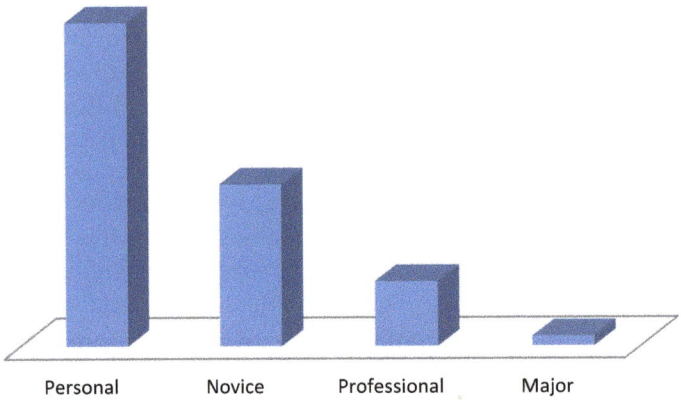

**Figure 1.1.** Relative frequencies of occurrence of kinds of creative thinking in physics that a student might engage in (reflecting the levels of creativity described by Beghetto and Kaufman (2009)). These categories are useful for thinking about our expectations, but, in reality, are rarely sharply defined or always entirely distinct.

## 1.4 Creative thinking doesn't come with a guarantee

Take the problem of explaining the tides. In 1616 Galileo had the idea that they were due to the Earth's spin and its annual journey around the Sun—the Moon had nothing to do with it. Today, the accepted idea is that tides are caused by the Moon and the Sun together. Ideas, even from illustrious scientists, don't come with a guarantee. In one sense, Dalton's, Thomson's, Rutherford's, and Bohr's models of the atom could be said to be 'wrong' (as might the quantum mechanical model in due course). These models, however, provided an understanding of the atom which fit the facts as they were known at the time, and allowed useful predictions to be made. But, creative ideas, when confronted by observation and tests, may have a shorter life than any of these. Becquerel's initial idea about penetrating radiation helps to illustrate this point.

### Becquerel and natural radioactivity

Henri Becquerel knew from his reading of Röntgen's work that cathode rays made the glass of a vacuum tube phosphoresce and emit penetrating radiation (see also chapter 2). In his mind he tied together the phosphorescence and the penetrating rays: if you get one, then you have the other. Given this notion he expected that a substance which fluoresces in visible light, such as a uranium compound, would also emit penetrating rays. Using Röntgen's approach, he tested the idea by putting a crystal of the substance on a lightproof cover over a photographic plate. Left in sunlight, it produced an image on the plate. This was consistent with the mental connection he had made. However, and to his surprise, when he left the crystal on a new photographic plate in a drawer, it also produced an image of the crystal. This made him re-think his simple connection of phosphorescence with penetrating rays. His new view was that the latter could exist without the former, and that the crystal itself was continually emitting penetrating rays.

Millikan's ingenuity led to a successful experiment that measured the charge on the electron. It seemed a plausible idea to apply the idea to measure the charge on quarks, but it failed. Does that make it a bad idea? A clever connection, notion, or idea that turns out to be a blind alley may be disappointing, but it can lead to a better idea provided that it prompts students to think again, rather than for them to persist with an unproductive line of thought, or simply give up. The point is that it matters less if their creative thought turns out to be 'right' or 'wrong', than that it should initially be plausible, tentative, and open to reconsideration. If students do this for themselves, all the better, but in the world of the physicist the reconsideration may come from a fresh mind. The story of Arago's disc shows this.

**Arago's disc**

Early in the nineteenth century François Arago found that rotating a copper disc above a compass needle made the latter rotate. At the time this was a real puzzle: how could a non-magnetic material (the disc) make a compass needle move? One suggestion was that the act of spinning the disc somehow made it magnetic. No-one less than Charles Babbage and William Herschel turned their minds to 'the magnetism manifested by various substances during the act of rotation' (1825 *Phil. Trans. R. Soc.* **115** 467). Their tests of various materials put them in order of susceptibility to the supposed effect. The list began with copper and zinc and ended with bismuth and wood, that is, it reflected the electrical conductivity of the materials. But their mindset was still on induced magnetism, rather than on magnetism inducing eddy currents. It was Michael Faraday whose different mindset saw the phenomena from two directions: if electricity could produce magnetism, could magnetism produce electricity? And, of course, this enabled him to construct the electromagnetic induction explanation that is now commonplace.

Teaching may skim over or ignore examples of creative thought of this nature. By focussing only on what is accepted today, what is currently held to be 'true' the lessons in them are lost. By focusing only on algorithmic rather than heuristic problems, the lesson may not even be there (Rogaten and Moneta 2016).

## 1.5 Variety, the spice of physics teaching

At this point we should, of course, bear in mind that a mini case study cannot do justice to the complexity of an historical event (Trefil 2002). It simply reminds readers of such events and of the possibility of using them as illustrative examples with their undergraduates. A particular physics course will have its particular foci so examples will reflect them.

A possible drawback of historical case-studies is that they come from a different world, one in which physics was almost exclusively the domain of men. A widespread gender imbalance still exists, partly because of socialisation differences between boys and girls and because of culture and upbringing differences, as in play.

Physics may still be seen as a masculine domain reflecting male interests (e.g. Cunningham 2013, Hazari and Potvin 2005) and, if we want more of a balance, we should be careful not to reinforce that perception. Students are attracted to physics for various reasons, and one kind of diet is unlikely to suit all. We should, therefore, also try to draw on the world today for examples and diverse interests, ranging from medical physics to astrophysics (Wulff *et al* 2018). Some aspects of what readers research themselves may also be useful examples which bring with them a sense of immediacy. Tutors may also reflect on the importance of role models for students, and on provision for academic and social integration to enhance the retention of under-represented groups of students (Strayhorn *et al* 2012). Opportunities for creative thinking add to the diversity offered by physics. Although widespread, we should note that self-selective participation in STEM subjects varies from country to country according to the culture.

We also need to be aware that while drawing on the more distant history of physics and Western achievement, it may make a point very well, but it can emphasise big events which some students may see as beyond their abilities. Paradigm-shifting ideas can make good stories, but the bread-and-butter physics which build on these can tell the same tale in an attractive way which reflects reality (Shneiderman *et al* 2006). Examples, drawn from a variety of contexts and times, can offset an overly romantic image some stories may encourage. At a time when there is concern for 'decolonising' what we teach (see also chapter 9), this opens the way to include more diverse examples from contemporary physics. There may also be a need to reflect on what students have been used to in their earlier education. When this has tended to have a strong, transmissive nature, students may feel uneasy about autonomy, expressing their own ideas, and uncertainty in the outcome, and press you for the 'right' answer. There may be an over-strong tendency to associate this with collective societies, or to those with, for example, a Confucian heritage. Ho (2020), in a very useful book on diversity and inclusion in global higher education, argues that it simply needs a sensitive response in teaching and it can be put to good use. For instance, so-called flipped teaching has become popular in recent times. Students study some aspect of a topic before they attend a session, and bring what they learn with them to explore it further. Ho recommends that the first part focuses on what has to be understood in the topic, then the second part can offer opportunities for small group, collaborative work of a creative, evaluative, or analytical nature. Such an approach shields students from a wider public failure, it frees them to be imaginative and try out ideas, and it allows the tutor to monitor and discretely offer suggestions. Approaches of this kind can be well suited to students from many cultural backgrounds. Moreover, diversity can bring new ideas and perspectives to collective creative thinking.

Diversity in teaching tools, as when having students use virtual labs, may similarly support the creative behaviours of some students (Gunawan *et al* 2018). (Such strategies are illustrated in later chapters.) This diversity has other benefits in that it widens the range of creative contexts brought to the students. Creative thinking and behaviour in physics has its own flavour, and it is more or less different to their flavours in other disciplines. At the same time, physics comprises various micro-domains and there can be subtle differences in the demands they make on creative thought and action. Baer and Kaufman (2005) use the analogy of the amusement park to make the point: it may be one amusement park but each 'amusement' calls for a different kind of engagement. Any variety we bring to students could make them aware that different physics contexts may make different intellectual demands.

## 1.6 Does it matter?

The fostering of creative thinking is important. One reason is the rapid advance in artificial intelligence and robotics. Many countries are turning their attention to the economic possibilities of a knowledge-producing economy. Competence in creative thinking will be at a premium in such a situation. As Rogaten and Moneta (2016) put it, creative thinking is 'essential to adaptation in a constantly changing work environment'. They add that little is done about it in higher education. Some creative competence can also convey a feeling of autonomy, an independence in thought and action, and a self-efficacy which can help to support well-being and the ability to cope in a rapidly changing world (e.g. Wright and Pascoe 2014).

More specifically, given that physics is a creative endeavour in its attempts to generate knowledge, it gives our students a fuller experience of what it means to be a part of that endeavour. Three attributes are considered to be advantageous in life: analytical abilities, practical abilities, and creative abilities (Sternberg and Lubart 1995). Courses may foster analysis and practical competence, but creative capability seems to receive little direct or explicit attention. It may be argued, of course, that creative competence is acquired by mere contact with the subject and those who teach it. It may also be that some may find it difficult to articulate what being creative in physics is—their creative thinking is just something that they do. In the twenty-first century, universities are likely to seek a more balanced approach to fostering the three abilities (Duderstadt 2018), and we feel that support for creative competence would be more effective if provision was at least conscious, deliberate, explicit, and thoughtful. Figure 1.2 illustrated just one way in which this might happen. For example, in the 'flipped learning' mentioned earlier, students study introductory materials themselves and then bring that knowledge and understanding to teaching sessions for further development at higher levels. Flipped learning is just one way of making space for other kinds of thinking, such as creative thinking (Ho 2020). This is, of course, one approach and an imaginative tutor will see others (see also chapters 8 and 9).

We do not, however, suggest that creative thinking can be transmitted to our students any more than understanding, critical thinking, or wisdom can be handed out. There is not a fixed procedure, algorithm, or formula for it, but we can make it more likely that these will develop through the experiences we provide, the teaching we do, and what we say. Like them, creative thinking can be practised to keep it alive, to keep it active, and to show it is valued. This practice also provides opportunities for noticing where students need more experience and what they are neglecting. A useful bonus is that opportunities for creative thinking allow students

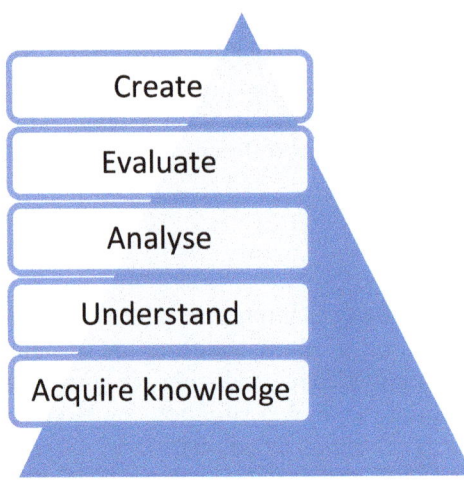

**Figure 1.2.** A well-known taxonomy of kinds of thinking, ranging from the 'lowest' level of thinking, the acquisition of factual kinds of information, to the 'highest' level, creative thinking (a version of Anderson and Krathwohl's adaptation of Bloom's taxonomy (Krathwohl 2002)).

to put something of themselves into their physics. This can be an attractive, motivating, and rewarding experience, and it may also support student retention.

We do not say that physics students everywhere are deprived of opportunities to exercise their creative thinking. These opportunities can, however, lack coherence and internal connection so that students get to play only fragments of the physics game. One perceptive student told us that such opportunities would be a welcome change from the tedium of mathematical exercises and laboratory report writing, and would offer interest and motivation. We would add that a coherent provision would teach them the nature of physics endeavour and help to prepare them for the world of scientific work.

We conclude, therefore, that exercising creative thinking in physics does matter. It gives students a fuller experience of what is, after all, their chosen subject, it supports the development of an increasingly important competence, and it can be motivating, satisfying, and attractive. The competence may not, however, develop without help, and may even decline if not exercised and valued. Development seems likely to be supported by:

- Raising students' awareness of creative activity in physics and of the wider value of creative competence.
- Increasing students' knowledge of the components of creative activity and its types in physics.
- Providing engaging opportunities to exercise creative endeavour, both individually and collaboratively.
- Encouraging and rewarding creative endeavour in its various forms.
- Providing constructive feedback about performance.
- Encouraging 'risk-taking' in thought, an acceptance that it doesn't come with a guarantee of an idea's success, and that 'failure' is also a learning opportunity.

- Modelling creative activity for the students.
- Maintaining a balance between creative activity and other kinds of activity in physics.

In giving students this experience of creativity in physics thinking there is, however, a need to guard against the culture of negativism which is common in the Western world. The aim is to encourage students' nascent attempts at creative thinking and develop a disposition to think creatively, not stifle it or destroy motivation. Many years ago, J B S Haldane (1963) wrote that the creative work of the professional scientist is generally judged:
- First, to be worthless;
- Then, interesting but perverse;
- Then, true but unimportant;
- And finally, 'I always said so'.

This can take a very long time. Perhaps we can be more generous with our students.

The next chapter selects and illustrates some fundamental ways in which the practice of physics is creative, and which lend themselves to student activity.

## 1.7 Something to reflect on

The practice of physics involves a variety of activities, some creative and some not.
- Readers might reflect on their personal experience of creative thinking in physics. Students may be engaged by accounts of this experience. Would any of it make a useful creative thinking case study for students?
- In practical terms, how might a historical report of a study in physics be used to illustrate creative thinking? What is the message you would want it to give to your students?

## References

Acar S, Burnett C and Cabra J F 2017 Ingredients of creativity *Creat. Res. J.* **29** 133–44
Arntz M, Gregory T and Zierhan U 2016 The risk of automation for jobs in OECD countries *OECD Social, Employment and Migration Working Paper* 189 OECD, Paris
Baer J and Kaufman J C 2005 Bridging generality and specificity: the Amusement Park Theoretical (APT) model of creativity *Roeper Rev.* **27** 158–63
Bakshi H, Frey C B and Osborne M 2015 *Creativity vs Robots* (London: Nesta)
Bao L and Koenig K 2019 Physics education research for 21st century learning *Discip. Interdiscip. Sci. Edu. Res.* **1** 2
Bates S P, Galloway R K, Lopston C and Slaughter K A 2011 How attitudes and beliefs about physics change from high school to faculty *Phys. Educ. Res.* **7** 020114
Beghetto R A and Kaufman J C 2009 Do we all have multicreative potential? *Math. Educ.* **41** 39–44
Beghetto R A and Kaufman J C 2015 Promise and pitfalls in differentiating amongst the Cs of creativity *Creat. Res. J.* **27** 240–1

Brodin E M 2018 The stifling silence around scholarly creativity in doctoral education *High. Educ.* **75** 655–73

Clark D 2020 *Artificial Intelligence for Learning* (London: Kogan Page)

Cunningham B A 2013 Women in physics *AIP Conf. Proc.* **1517** 173–87

deWitt J, Archer L and Moote J 2019 15/16-year-old students' reasons for choosing and not choosing physics at A level *Int. J. Sci. Math. Educ.* **17** 1071–87

Duderstadt J J 2018 Preparing the American university for 2030 *The Future of the University in a Polarizing World* ed L E Weber and H Newby (Geneva: Association Glion Colloquium), pp 193–204

Egan A, Maguire R, Christopher L and Rooney E 2017 Developing creativity in higher education for 21st century learners *Int. J. Educ. Res.* **82** 21–7

Frey C B and Osborne M A 2013 *The Future of Employment* (Oxford: Oxford Martin Programme on Technology and Employment)

Gunawan G, Suranti N M Y, Nisrina N, Herayanti L and Rahmatiah R 2018 The effect of virtual lab and gender toward students' creativity of physics in senior high school *J. Phys. Conf. Ser.* **1108** 012043

Haldane J B S 1963 Review of 'The Truth about Death' *J. Genet.* **58** 464

Hazari Z and Potvin G 2005 Views on female under-representation in physics *Electron. J. Res. Sci. Math. Educ.* **10** 1–33

Ho S 2020 Culture and learning: Confucian heritage learners, social-oriented achievement and innovative pedagogies *Diversity and Inclusion: Global Higher Education* ed C S Sanger and N W Gleason (Singapore: Palgrave Macmillan), pp 117–62

Jackson N and Shaw M 2006 Developing subject perspectives in creativity in higher education *Developing Creativity in Higher Education* ed N J Jackson (London: Routledge-Falmer), pp 89–108

Jahnke I, Liebscher T and Wildt J 2017 Teachers' conceptions of student creativity in higher education *Innov. Educ. Teach. Int.* **54** 87–95

Kaufman J C and Beghetto R A 2009 Beyond big and little: the four C model of creativity *Rev. Gen. Psychol.* **13** 1–12

Krathwohl D 2002 A revision of Bloom's taxonomy *Theory Pract.* **41** 212–8

Lebedeva N, Schwartz S H, Van De Vijver F J, Plucker J and Bushina E 2019 Domains of everyday creativity and personal values *Front. Psychol.* **9** 2681

MacLaren I 2012 The contradictions of policy and practice: creativity in higher education *London Rev. Educ.* **10** 159–72

Marquis E, Radan K and Liu A 2017 A present absence: undergraduate course outlines and the development of student creativity across disciplines *Teach. High. Educ.* **22** 222–38

Ramge T 2019 *Who's afraid of AI?* (New York: The Experiment)

Rogaten J and Moneta G B 2016 Creativity in higher education *Psychology of Creativity* ed G B Moneta and J Rotagen (New York: Hauppange), pp 3–20

Shneiderman B et al 2006 Creativity support tools *Int. J. Hum.-Comput. Interact.* **20** 61–77

Simonton D K 2012 Taking the US patent office criteria seriously *Creat. Res. J.* **24** 97–106

Sternberg R J and Lubart T I 1995 *Defying the Crowd* (New York: Free Press)

Strayhorn T L, DeVita J M and Blakewood A M 2012 Broadening participation among women and racial/ethnic minorities in science *Social Inclusion and Higher Education* ed T N Basil and S Tomlinson (Bristol: Policy Press), pp 65–82

Trefil J 2002 *Cassell's Laws of Nature* (London: Cassell)

Wilczek F 2016 How Feynman diagrams almost saved space *Quanta Magazine* www.quantamagazine.org/20160705-feynman-diagrams-nature-of-empty-space (Accessed: 27 June 2021)

Wright P R and Pascoe R 2014 Eudaimonia and creativity *Camb. J. Educ.* **45** 295–306

Wulff P, Hazari Z, Petersen S and Neumann K 2018 Engaging young women in physics *Phys. Rev. Phys. Educ. Res.* **14** 020113

**IOP** Publishing

Creative Thinking in University Physics Education

Douglas P Newton, Sam Nolan and Simon Rees

# Chapter 2

## The creative learner in physics

### 2.1 Learning physics and learning what counts in physics

Creative thinking in physics is more or less **novel** (imaginative, original, at least to the student), **appropriate** (plausible, fit for purpose), and preferably in some way **impressive** or **satisfying** (clever, concise). But where are the opportunities for thought that is imaginative, plausible, and clever? The Doppler effect provides a useful case study with meaningful messages for students.

#### The Doppler effect

The development of relatively fast steam trains in the nineteenth century brought a phenomenon to people's attention. As trains approached and receded, they sounded different. There was a pitch change, higher as they approached, and lower as they receded. Christian Johann Doppler (1803–53) with an interest in wave theory, explained what became known as the Doppler effect. The wavelength of the sound was 'compressed' as the train approached and 'stretched' as it receded. Previously, Doppler has predicted that would happen with star light but lacked a means of testing the prediction. However, in the case of sound he asked C B Ballot to test it. Ballot's test of 1845 was remarkable for its ingenuity and its colourful nature. He had trumpeters play the same note on an open car of a train while, at the track side, he had other trumpeters play the same note. As the train approached and receded at a known, constant speed, the differences between the notes agreed with Doppler's calculations.

Sound waves, of course, are not like light waves and the effect had not been confirmed for the latter. However, A H L Fizeau was able to confirm it using shifts he observed in the spectral patterns of star light. The Doppler effect predicts that the wavelength of light will be shifted towards the red end of the spectrum as the emitter recedes. Edwin Hubble, observing this red shift in the spectra of galaxies, showed that galaxies are moving apart and as the distance increases it does so at increasing rates. In short, it led to the idea of an expanding Universe and the so-called Big Bang theory.

This points to some major opportunities where creative thinking is a valuable tool. Put simply, Doppler had **noticed the opportunity** to extend wave theory to moving sources. His mind was **prepared** for the 'new' phenomenon presented by the sound of moving trains and he **noticed the connection** with his interest. Second, he **constructed an explanation** of effects of this kind. Third, **imagination** produced tests of that explanation, first for sound, then for light. Fourth, the explanation was **applied in another area** to support understanding there. But there is another message in the story—it does not all have to stem from the creative thinking of one person, it can be a **joint enterprise** which extends over time. The problem of the nature of matter, outlined in the previous chapter, also shows that problems can be very long-standing and may be revisited and solved differently over time.

Summing this up, there are major opportunities in:
- **Noticing**, making sense of, and formulating a problem, puzzle, or opportunity.
- **Constructing** a tentative explanation, idea, or a model which can be articulated to make a prediction.
- **Testing** the prediction of the tentative explanation, idea, or model and developing it further.
- **Applying** the explanation or idea.

These may also involve, for instance, **generating** new concepts, perspectives, tools (mental and physical), and approaches to measurement.

This may give the impression that creative thinking is always part of a highly organised and step-by-step process in which the physicist is first faced with a puzzle and then generates some kind of tentative solution, followed by tests of it, and then uses it or offers it in applications elsewhere. In reality, each of these opportunities is itself complex, and the path is not always, perhaps rarely, smooth from the first to the last. There are often false starts which necessitate a return to an earlier stage in any part of the process. The process is an iterative one and luck can play a part too, and not everyone takes it through from noticing to applying themselves. Some may find their interests and skills lend themselves better to, for instance, theoretical work, while others prefer experimental activity, or the application of knowledge to solve a particular problem. Some may also prefer to visualise and articulate mental models of the world, while others are more interested in following and interpreting the mathematics, and their creative thinking will reflect this (Dirac 1982).

## 2.2 Noticing and making sense of problems

Questions can arise from observed regularities. Isaac Newton said that he noticed an apple falling from a tree and that prompted him to ask why it fell down and not up or sideways. It set a chain of thought in motion which generalised from the apple, to all matter being attracted to matter in proportion to their masses, and to its relevance to the motion of planets. But people had seen apples fall from trees every year and hardly gave it a thought. It might have been the same for Newton had his mind not been tuned and prepared to see this mundane event from a different perspective. And not every interesting puzzle is of momentous consequence. For instance, take the case of Cronquist's broken-off pencil points.

### BOPPs: a case of noticing and wanting to know why

Cronquist reported that one day, as he cleared his desk top, he noticed 'a very large number of broken-off pencil points (BOPPs) lying between and behind the books' on his desk. More significantly, he noticed that these BOPPs were almost all identical in size and shape. Why was that? His stress analysis showed that there was a place along the lead where the tensile stress during writing was at its greatest, and this was determined only by the size and shape of the pencil point. He found that points produced by the same sharpener (and so having the same size and shape) would break where the diameter of the cross section is 1½ times the diameter of the pencil point tip. Hence, his desk was scattered with BOPPs of the same size and shape. Had he been able to use a pencil until it became quite blunt, the BOPPs would have been longer because the tip diameter would have been greater. Presumably, he was somewhat heavy handed and broke the points soon after sharpening the pencils.

Such a study (reported in *American Journal of Physics* **47** 653 in 1979, and again online in 1998 at https://doi.org/10.1119/1.11954) is unlikely to attract fame or fortune. Perhaps many such pursuits of curiosity never go beyond the desk of those who follow them. But the noticing, the 'Why?' question it generates, and the urge to know the reason, is often where creativity starts.

Not every puzzle, however, is noticed by the one who tries to solve it. For example, the decay of a neutron into a proton and an electron seemed to go against the principle of the conservation of energy. How could this be? In 1930 Wolfgang Pauli's tentative solution was that a 'particle' of zero mass and no charge was needed to balance the energy equation. The existence of this particle, later called a neutrino, was confirmed by experiment in 1956. Some students may think that the regularities of nature (the laws of physics) are the end points of research, when they are often only the beginning. Take, for instance, the heat emitted by a black body. Joseph Stefan discovered that it was proportional to the fourth power of the temperature. Only later was a theoretical understanding developed (by Ludwig Bolzmann) after the discovery of the fourth power law. Even something seemingly as simple as Snell's laws of refraction can beg the question. Why does a light ray change direction when it enters another medium? Similarly, an interesting phenomenon such as super-conductivity may be explored to find its limits but it needed an explanation, provided initially by Cooper's electron pairs but remaining to be explained at higher temperatures. And some problems have long histories. Heinrich Olber (and others) pointed to a problem with the long-standing belief that the Universe is without limit and contains an infinite number of stars. Olber's question was that, if this is so, why is the night sky dark? This was not answered convincingly until the twentieth century with, amongst other ideas, the construction of the Big Bang theory. And some questions are fundamental, of enormous consequence, and seemingly unassailable, like 'Why is there something, rather than nothing?

Sometimes, the question, puzzle, or problem that starts the process is not always the one that is answered at the end. But this often needs a prepared mind, one which already holds relevant information so that a chance event is noticed, seen as more than trivial, and a new question is formulated and pursued. Louis Pasteur said that 'chance favours the prepared mind'. This short vignette about the discovery of x-rays illustrates the roles of serendipity and the prepared mind.

### The discovery of x-rays

Wilhelm Röntgen (1845–1923), interested in the penetrating abilities of cathode rays in discharge tubes, repeated some of the investigations of others to become familiar with the phenomenon. While such rays had been observed to pass through thin materials, such as aluminium foil, no-one had observed them passing through the glass walls of the tube itself. Could it be that they could do so but the glow of the tube swamped the fluorescence of the detecting screen? To answer the question, Röntgen covered the tube with opaque card, darkened the room, and examined the tube for leakage of light. Satisfied, he was about to turn on the light when he noticed that his detecting screen lying some distance away was fluorescing. It was well away from and out of the direct line of cathode rays in the tube. At first, he thought some light might be escaping from the tube, but his checks assured him it was not. This suggested that it was not cathode rays passing through the glass, but was something produced by the discharge that was able to pass through the card cover and travel some distance in air. The effect showed no response to electrical and magnetic fields so was not due to charged particles. Unsure of what he had noticed, he called the radiation x-rays and explored their properties over several weeks of intense work, culminating in the well-known x-ray image of his wife's hand on a photographic plate. It would have been easy to overlook the fluorescence on the screen, or dismiss it as of no consequence. An informed, prepared mind, however, saw that there was potentially some significance in the unexpected effect.

As far as students are concerned, few may have had the opportunity to find their own problems to explore, even theoretically. As they observe the physical world, questions may arise in their minds which remain unanswered. Some will be beyond their experience, skills, ability and means, but some, like BOPPs, may be open to further inspection.

Noticing problems is only the beginning. Often, they are problems because they are looked at from an unproductive perspective or they are in a form which does not lend itself to being solved. They often need to be reformulated, simplified, taken to their bare essentials, and rendered amenable to solution. This reformulation takes imagination and a 'What if?' mode of thought: What if I looked at it in this way? What if I simplified it in this way? What if it is really more than one problem? A combination of analysis (asking 'What does this mean?') and reformulation is generally needed to dismantle and reconstruct a problem. The process itself may also generate other problems.

## 2.3 Constructing understandings to enable explanation

As a species we try to make the world meaningful by making mental connections between events and their contexts. This has some survival value. Particularly powerful connections are those that link cause and effect as these make the world predictable and provide a working understanding of it. Often, we expect students to make connections like those in the understandings created by others. We can't, of course, force students to make the connections—they have to do that for themselves. However, we often try to highlight what matters in some way in order to increase the likelihood that they and the relationship will be noticed. This is a routine activity of teaching (Newton 2012). In physics it can make matters meaningful, understandable, predictable, and applicable in new situations.

The making of new connections underpins creative thinking (Arthur Koestler (1964) called this 'bisociation', the bringing together of pieces of previously unconnected knowledge). The aim is to have students use this natural tendency to create new understandings themselves. They must explore a problem and make connections with prior or newly acquired knowledge more or less unaided. Bisociation is about meaningfully connecting two pieces of knowledge, but, of course, the process can go beyond that to produce quite complex structures meant to represent and function like some aspect of the world. The models of the atom described in chapter 1 illustrate the creative construction of such mental models. One such mental model which has become a long-standing mainstay of scientific thinking is the mechanical model of a gas created by Daniel Bernoulli (1700–82), now known as the kinetic theory of gases.

### The kinetic theory of gases

Bernoulli's clever idea was to visualise a gas as comprising particles in random motion. These particles bombard the surfaces around them and so produce the effect of pressure on those surfaces. Heat makes the particles move faster and that increases the pressure. Compression confines them in a smaller volume so more bombard a given surface area, again producing more pressure. The random motion of the particles results in other effects, such as gaseous diffusion. Bernoulli's model can be readily imagined and mentally manipulated to predict and understand such phenomena. James Clerk Maxwell (1831–79) later gave this qualitative model a mathematical basis which enabled it to have a more precise form and lend itself to detailed explanation and prediction, but the simple model is now commonly taught in schools.

Of course, connections may be made between anything in a student's mind to produce a very large number of understandings. They are subject, however, to the constraint of plausibility: the more plausible a connection, the more acceptable it is until it is put to the test. Given equally plausible explanations, it is also usual to apply Ockham's razor and cut out the more complex explanations in favour of the simplest (until proven otherwise), a principle due to William of Ockham (or Occam) (1285–1349). And, if it is somehow satisfying or clever, it adds value to the idea.

For students, a note on a distinction between a potential explanatory idea and a theory could be instructive. In science a theory is an explanatory idea that has been successfully developed and tested and goes beyond existing explanations, preferably through predictions which distinguish it from them. An example of an explanatory idea is Stephen Wolfram's 'theory of everything', a computational approach aimed at explaining the Universe. It shows how complex structures can develop by applying a few code-like rules to a network of connected points. A number of physicists object to this being described as a 'theory' without evidence that the idea at least equals what already exists and offers predictions which can distinguish it from them (see Becker (2020) for a student-friendly account). On the other hand, 'the kinetic theory of gases' meets these expectations. Students' speculative ideas, however, are unlikely to be on this scale, and more often will draw upon or apply established theories to phenomena, in accordance with the expectations depicted in figure 1.1. And, of course, not all ideas are to do with explanations.

## 2.4 Testing a tentative explanation or idea

The testing of ideas often involves a specific articulation of an idea that lends itself to the construction or assemblage of tools and equipment which enable observation and measurement. For instance, Maxwell's equations predicted the existence of long wavelength electromagnetic waves ('radio' waves), but no-one had been able to produce them. Heinrich Hertz took on the challenge. He generated electrical sparks across a gap between two metal balls, something that was commonly done at the time in demonstrations of static electricity. But in his case, he also set up a loop of wire with two metal balls similar to those in the spark generator, but several metres away. When the spark was produced by the generator he saw a small spark appear in the wire loop. He concluded that electromagnetic waves of about 3 m in length had been emitted by the spark generator and had reached the loop, generating a spark there. This was evidence consistent with the predictions of Maxwell's equations. Marconi, of course, was the one who turned it to commercial advantage.

Where is the innovative thinking in this? The Maxwell equations say nothing about the specific ways in which such waves can be produced. Moving out from those equations needs a mental leap to connect it to the tangible world of equipment and measurement (or vice versa). Hertz made that leap, connecting the possibilities of the spark generator with the promises of the equations. In addition, he also had to devise a 'detector', the loop of wire, and the means of measuring the wavelength. But, imagination, even of the most ingenious person, can have its limits.

Testing such ideas is not always easy. Considerable creative effort and ingenuity may be needed to produce a test which most accept as fit for purpose. For example, experience prompts us to believe that a medium is needed for wave motion. If so, how can light travel through empty space? To solve the puzzle, the ether was proposed, an insubstantial material which fills space and provided a medium for light waves. Did this really exist? Albert Michelson and Edward Morley constructed a classical example of an ingenious test of a reasonably plausible idea.

### The ether and ingenuity

Earth travels through space at about $3 \times 10^4$ m s$^{-1}$, and light travels through the ether at some $3 \times 10^8$ m s$^{-1}$. Assuming that the ether exists, drawing a parallel with other kinds of wave, Michelson and Morley assumed that when the Earth approached a light wave through the ether, the light should appear to be travelling slightly faster. Conversely, when the Earth receded from a light wave, the light should appear to travel slightly slower. Because the Earth travels so much slower than light, the change in speed is difficult to detect. The experiment they devised involved splitting a beam of light and sending the halves at right angles to one another to be reflected back and combined. Michelson, already skilled at making optical devices and in measuring the speed of light, used an interferometer to view the returning light. If there was a difference in speed, even small, the interference pattern should show it. Neither sending one half of the split beam with the Earth's motion nor sending it against that motion produced a change in the speed of light. The notion of the ether had to be rejected.

Where is the creativity in this experiment? Had these been sound waves, perhaps they would have measured the speed of sound emitted from a train as the train approached and receded. In the case of light, a train would need to move at an impossible speed to make a difference. Michelson and Morley chose the fastest object easily available, the Earth. Even with this, the task of measuring the change in speed would be difficult. Michelson brought his prior experience of interferometry and the measurement of the speed of light to the task. The final, clever idea was to split a light beam and send it in two directions at right angles to one another. It all makes a good case study for students.

The notion of the ether had to be rejected, but investigations can go the other way and provide evidence consistent with predictions, as it did with Hertz evidence for the existence of Maxwellian electromagnetic waves ('radio' waves). Of course, practical investigations can serve a variety of purposes. Millikan's oil drop experiment, for instance, was to determine the likely charge of an electron. He was not testing the existence of the electron or that it was negatively charged. Instead, he created an ingenious, clever way of determining its probable charge, something of enormous value because it facilitated a wide range of other activities.

## 2.5 Application

The application of tried and tested ideas and concepts is manifold. For example, one application is in the understanding of events in the everyday world, away from the laboratory. Helen Czersky, physicist and science broadcaster describes such an event on her way to a documentary location (Czersky 2015). The cameraman opened the car window a little to take pictures. This produced a low-pitched, loud noise in the car which Czersky recognised as an instance of the Helmholtz resonator, like the higher-pitched sound produced by blowing across the top of a bottle. This was confirmed when a second window was opened and the car interior no longer functioned as a resonator. Here, an idea in physics was applied to a naturally occurring event in the everyday world and rendered it agreeably meaningful. Some events involve our own actions for more practical purposes. For instance, the Fosbury Flop is an approach used in the high jump in athletics. The athlete jumps, curls backwards and rotates over the bar in the shape of a letter C with arms and legs pointing to the ground. This shape puts the centre of gravity of the athlete in the hollow of the C. Although the body clears the bar, the centre of gravity passes *under* it. A greater clearance in achieved for a given input of energy. We should not underestimate the interest and satisfaction this kind of application has for our students, many of whom will not become professional physicists but will find the world more understandable through physics. Applying physics to construct understandings of events in the everyday world is a reasonable creative expectation we might have of our students.

Another application is in the understanding of other phenomena of interest in physics. For instance, Czersky (2011) drew on accepted ideas to extend her understanding of bubbles. She knew that when two bubbles coalesce, a sound is emitted at the natural frequency of the new bubble. But what mechanism generates the pulse of sound? Making some simplifying assumptions, and drawing on prior ideas and data about bubble fusion and fragmentation, the amplitude of the sound was in line with the effects of the rapid increase in bubble volume, and, moreover, depended on the size of the smaller, 'parent' bubble. There are many well-known examples of this kind of application as normal physics builds upon successive, successful ideas. For example, Max Planck explained the energy distribution emitted by a black body by assuming that electromagnetic radiation comprised quanta of energy. Einstein explained the photoelectric effect by invoking the same notion for light (an explanation ingeniously tested by Millikan of oil drop experiment fame). This kind of application and building only tends to be disrupted at times of what Kuhn (2012) described as major and fundamental revolutions in thinking, but then it starts again as the dust settles.

A third kind of creative application is in the construction of a tool to use in practical investigations. For instance, most are familiar with the Bunsen burner of school science. But Robert Bunsen's name isn't generally linked with the spectroscope which he invented using a prism in a cigar box fitted with telescope lenses. (His burner was the tool he needed to raise the temperature of substances high enough to show their characteristic lines in his spectroscope.) A similar application is when an accepted idea is used to solve practical problems like those met in technology. For instance, in medical physics, the rate of flow of urine can inform judgements about a progressively increasing obstruction. A simple and cheap way of measuring the rate of flow is to use a funnel-shaped device with a small hole in the bottom. The liquid rises in the funnel until the rate at which it flows in matches the rate at which it runs out. In that steady state, Bernoulli's theorem can be applied to predict the relationship between the rate of flow and the height of liquid in the tube, the latter becoming an indicator of the volume per second entering the funnel. Such a simple device can be used anywhere in the world, even where medical equipment is basic (Newton 2008).

## 2.6 Creative thinking is not a mechanical process

Each of these kinds of creative thinking can be more or less novel, more or less appropriate, and more or less clever in its context. In addition, these fundamental aspects of physics often also rest on other innovative ideas and concepts. The notion of a black body in physics, for instance, refers to a hypothetical object which absorbs all heat and light that falls on it. It has been a useful concept for developing a detailed, theoretical understanding of heat and light emission from objects in diverse contexts. Thinking creatively is central to what a physicist does, and in the workplace it may extend to associated needs and problems, such as the use of available equipment, the organisation of a team, the division of labour, and the construction of a bid for funds. While these may materially contribute to the feasibility and success of a project, they may not themselves determine the scientific thinking in it. Nevertheless, a postgraduate who aspires to university life will, in due course, need to learn of such things. Such deliberations take time—a creative solution does not always come quickly, or even when thinking about physics. For instance, physicists have reported that some 20% of their 'significant' ideas had their origins in non-work related activity (often referred to as a period of incubation), or during mind-wandering. Working at CERN in Geneva, Michael Deveaux wanted to test the use of smartphone cameras as hadron detectors, but he urgently needed a moveable phone carrier to place in the particle beam. When playing with his two-year-old daughter, he had the idea that he could use Lego™ bricks to build the carrier, and it worked, although he had to add a sign which pointed out, 'This is not a toy!' (Rao 2020). At times, a break from a difficult problem can also let an unproductive approach or mental block fade and free the mind to think in another way (Gable *et al* 2019).

It is important to emphasise again that the creative process can be messy, iterative, disordered, and sometimes disconnected. For instance, there was a prolonged debate about the Boltzman's mathematical interpretation of the kinetic theory of a gas: did such particles really exist in a gas, or were they simply a mathematical convenience? Boltzman found the dispute was so disheartening that it contributed to his suicide. Unluckily, Robert Brown's observation of the random motion of pollen grains had been neglected for almost a century until Einstein picked it up and saw it as evidence of the reality of these mathematically convenient particles. A more recent example is offered by the familiar notion of dark matter. Dark matter was postulated to make up for an apparent deficit of other matter in the Universe. Vera Rubin, in the 1970s, explained the movement of hydrogen atoms in galaxies as being due to the influence of such matter. But, the question remains: if there is dark matter, what is it?

Here, we have given the creative process a tight order for the sake of simplicity and emphasis. Students also need to appreciate that it is not unusual to be creative and wrong (and being wrong can be a part of the learning process experienced by professional physicists). In addition, there is no algorithmic procedure which guarantees successful, creative outcomes, and that the process is not always smooth or entirely systematic. The way such creative activity is presented in research reports and textbooks, however, often hides the sweat, tears, false starts, and the steps from problem sensing to persuasion that went into it. For instance, objects (even Lego™ bricks) may be used to represent, communicate, and develop abstract ideas. Such objects become 'tools for thought' and make concrete what is being said, opening an idea for articulation and elaboration (Bjørndahl *et al* 2014). New knowledge tends to be presented in tidy, dispassionate accounts, without the details of the pathway that led to it, and not as a human endeavour. And this is not a bad thing for a subject that strives to be as objective as possible about its outputs or products. But it tells us (and our students) little about *how* those products came into being. When teaching, we have to put something of the people and the process back in to do that. There are rewards, not least being the benefits of engagement, particularly for those who have previously seen the subject as boring and unattractive. Finally, the process of working in physics has been described as a kind of exploratory, constructive play, and an inclination to play in this way has been seen as a newcomer's passport to acceptance as a physicist (Hasse 2008).

## 2.7 Something to reflect on

- Students' notions of what counts in a subject tend to be shaped by the tasks they are given. What hidden messages could be in the tasks given to the physics undergraduates you know?
- A *bridging course* aims to take students from one kind of thinking or way of working to another. If students believe that understandings in physics are not products of creative minds, could a short bridging course help them see the subject as more 'dynamic' and creative? What might such a course look like?

## References

Becker A 2020 Physicists criticize Stephen Wolfram's 'theory of everything' *Sci. Am.* **322** 5 https://scientificamerican.com/article/physicists-criticize-stephen-wolframs-theory-of-everything/ (Accessed: 5 July 2021)

Bjørndahl J S, Fusaroti R, Østergaard S and Tylénk K 2014 Thinking together with material representations *Cogn. Semiot.* **7** 103–23

Czersky H 2011 A candidate mechanism for exciting sound during bubble coalescence *J. Acoust. Soc. Am.* **129** EL83–8

Czersky H 2015 How to make music while driving through a desert *Sci. Focus* 284 27 https://www.sciencefocus.com/science/everyday-science-with-helen-czerski-how-to-make-music-while-driving-through-a-desert/ (Accessed: 7 December 2021)

Dirac P A M 1982 Pretty mathematics: playing with the maths *Int. J. Theor. Phys.* **21** 603

Gable S L, Hopper E A and Schooler J W 2019 When the Muses strike: creative ideas of physicists and writers routinely occur during mind wandering *Psychol. Sci.* **30** 306–404

Hasse C 2008 Learning and transition in a culture of playful physicists *Eur. J. Psychol. Educ.* **23** 149–64

Koestler A 1964 *The Act of Creation* (New York: Penguin)

Kuhn T 2012 *International Encyclopedia of Unified Science. Part 2: The Structure of Scientific Revolutions* ed O Neurath vol 2 (Chicago, IL: University of Chicago Press)

Newton D P 2008 The uflow meter: a peak flow measurement device *Patent* GB 241183

Newton D P 2012 *Teaching for Understanding* (London: Routledge)

Rao A 2020 Using Lego to study the building blocks of the Universe *CERN* 7 Feb https://home.cern/news/news/experiments/using-lego-study-building-blocks-universe (Accessed: 27 June 2021)

IOP Publishing

Creative Thinking in University Physics Education

Douglas P Newton, Sam Nolan and Simon Rees

# Chapter 3

## Creative thinking in practice: problems

### 3.1 Fertile problems

If students are to exercise their creative thought they need something to think about, what we usually refer to as a 'problem'. Unfortunately, there are many things we have students do that we call problems and some do not provide much opportunity for creative thinking. This is not to say that such problems are worthless. For example, students may solve problems which lead to particular (already known) outcomes, generally produced by manipulating the mathematics. In the process, their grasp of the topic may be improved. Here, we focus mainly on those problems which ask questions such as 'Why?' and call for an explanation. These can be fertile starting points in that they allow space for imagination, and the outcomes can lead to further creative thinking. As a kind of question used in teaching, they can be overlooked and neglected, sometimes because they can lead to unanticipated or unexpected ideas. Nevertheless, these questions begin the process of investigation in physics, as in other sciences.

### 3.2 Curiosity and questions

Perhaps when you start a new topic you have a lot of questions from your students. But are they the questions which might lead to creative thinking? Probably not, at least at the beginning. If the topic really is new to the students, it's likely that their first questions (and their second questions and third questions...) will be about matters of fact. In the early years of the twentieth century, when electrical energy was just beginning to be important in industry and everyday life, matters of fact about the older subject of static electricity were still being collected. For example, Hertha Ayrton (1854–1923), in a letter to *Nature* (17 February, 1902, p 390) wrote about what she did. She described how a candle flame placed up to 40 cm from a charged electroscope made it lose its charge. This happened faster when the candle was earthed but still happened when it was insulated, albeit more slowly. Even the flame of a match and a red hot wire had this effect. But she noted that what was especially interesting was the effect of cold bodies. Just as effective as the candle flame was 'cotton wool dipped in ether, methylated spirit, or diluted sulphuric acid', each placed on an insulated saucer. She pointed out that this observation was something that had not previously been reported, and concluded that, 'Many other experiments were tried, but the preceding are sufficient to show the nature of the phenomena observed'. Curiosity-driven exploration of an effect accumulates facts, information, and data about it which enables a **descriptive** mental model to be constructed. Eventually, this may also include incomplete or simple ideas about causes, perhaps only partly conscious at this stage, and it can guide further questioning, observation and investigation. This may be what prompted Ayrton to try cold liquids which can evaporate and spread through the air to the electroscope. At some point, the question moves from 'What?' to 'What if …?', and to a conscious asking of 'Why?', and to the formation of an embryonic explanation. The descriptive mental model is now an **explanatory** model which includes a cause, and may be articulated to make predictions that are potentially open to testing.

It may take a number of factual questions before a student develops a feel for the subject, begins to ask 'Why', and then seeks a causal explanation. Such questions are rarely expressed in this way in today's scientific reports, but instead are disguised in somewhat sterile statements of aims. Past physicists seemed more at ease with them. In a paper describing her investigations of the origins and growth of ripple marks, Hertha Ayrton (1910) began, 'To any one who, for the first time, sees a great stretch of sandy shore covered with innumerable ridges and furrows, as if combed with a giant comb, a dozen questions must immediately present themselves. Are they made and wiped out with every tide, or do they take a long time to grow and last many tides? What is the relation between the ripples and the waves to which they owe their existence?' Aytron's reformulation of her questions led her to focus first on the cause of the ripples. With students, such questions can set the ball rolling and, if you don't provide a ready answer, can prompt students to think for themselves, refine their questions, explore the phenomena, and generate their own tentative explanations. Impatience, however, can stall the process so that students abandon any interest in generating a tentative causal explanation which could underpin subsequent creative work. However, while we generally welcome 'Why' questions as our starting points, we should be ready for those students who need to lay the factual foundations first. Depending on the topic, this may need time for exploration and reflection, so lower order questions are likely to be asked before higher order questions (Aflalo 2018).

## 3.3 Noticing, finding, and posing problems

A question without an obvious answer begins the creative process in physics. The physicist finds questions like this in various ways. For instance, an existing explanation may fail in the light of an observation, curiosity about some phenomenon may prompt thought about an explanation, extending an analogy may lead to 'I wonder if …' reflections, or new tools renew interest in long-standing puzzles (Kim *et al* 2012). Each has the potential to generate a problem.

In teaching physics, however, the term 'problem' may be used rather loosely to include exercises set for students which can be successfully completed by following examples, rules, established procedures, or algorithms. There can be a tendency to overdo the well-defined problem to the exclusion of others (Williams 2018). Perhaps as a consequence, there are students who expect to solve all problems simply by picking up the 'right' formula, putting numbers in it, and mechanically processing it. Many real-world problems, however, are often ill-defined and complex, and solutions are rarely found by following a script. Procedures, formulae, and algorithms can help, but by themselves they rarely point to a clear path to the solution. The uncertain process of finding a path 'without a script' is what makes it a problem and experience of that process is a lesson for the student (Martinez 1998).

At this point, for teaching purposes, it is useful to distinguish between problems that are *noticed*, those that are *found*, and those that are *posed*. According to Ayrton's account she, herself, *noticed* a need to account for the appearance of ripple marks in the sand. Cronquist similarly noticed the regularity of broken-off pencil points (chapter 2) and wondered 'Why?' It also includes, for example, noticing the need for a concept, a term, a representation, the construction of a practical test, or an opportunity for an application. Here, noticing indicates that these are products of direct or vicarious experience of an event, phenomenon, or related information. Problem *finding*, on the other hand, may be defined as the outcome of a successful search of sources in which the problem is stated or embedded. For instance, in China Qi and Zhang (2020) were aware of the well-known problem with present measures of Hubble's constant—they produce different results. This prompted them to propose and explore the potential of black hole shadows to provide another measure. Here, the distinction between a noticed and a found problem is in their origin: the former is a first-hand experience, while the latter is second-hand (without attaching value judgements to these terms). There are also problems that are *posed* or given to students to solve. These have generally been collected, constructed or selected by a tutor and usually relate directly to some aspect of the physics being taught. They may comprise problems noticed by past students or the tutor, or problems found by the tutor from other sources. For example, a tutor would know that a recent scrutiny of data about the mass of the W boson does not match predictions from the Standard Model of elementary forces and components of matter—in short, there is a problem (Aatonen *et al* 2022). Posing problems may, at times, exercise the tutor more than the students, but in this context it is useful for practising problem analysis and reformulation, important capabilities in creative problem solving.

We could, however, ask students to draw on their own mental resources and *notice* for themselves a physics problem or puzzle in the world around them. ('Mind wandering' around an observation—thought that is only loosely focused and drifts in and around it—can help to construct such puzzles (Irving 2016).) The noticing of a physics problem, like Ayrton and Cronquist did, may lead to a feasible investigation. Its solution may not be entirely new to the world, or it may involve the application of an existing theory, but self-generated problems can be motivating and interesting for the student and can lead to what has been termed 'curiosity-based learning' (Savery 2006).

Perhaps what is surprising is that such problems may have gone unnoticed or, at best, have been seen as natural regularities of little interest. Nevertheless, even unambitious problems may be difficult to accommodate in a course. What students notice may not relate directly to the physics being taught and resources may not be readily available to support it. However, with some constraints and guidelines for the students, noticed problems may become mini-projects with a place in a short course dedicated to raising an understanding of the role of creativity in physics.

Alternatively, we might ask students to *find* a physics problem in some topic. Generally, this will have been noticed and recorded by practising physicists. Problems relating to current interests in physics have often been highlighted and await solutions, and scientists often work on problems which they did not, themselves, notice. For instance, Ayrton's problem has arisen again in a very different form. *Nature* (29 April 2021) reports that Voyager 1, which left the Solar System in 2012, has detected ripples in the electrical charge of the interstellar plasma. What is their cause? While problem noticing calls for observation of and reflection on the real world, problem finding often involves interaction with the literature or conversations with practitioners. This points to another kind of activity for students which adds to their experience of physics as an endeavour. As the teaching of a particular topic progresses, the students may be asked to explore the literature on it and delineate the limits of our knowledge by giving particular attention to recognised and, as yet, unsolved problems. More experienced students may be asked to take this further by selecting a problem and reflecting on what its solution would entail. Found problems are likely to be challenging. Candidates for solutions could lead to something new to the world (and, perhaps, even be an instance of professional creativity, as described in chapter 1), but may also be beyond the resources, skills, and experience of undergraduate work. Nevertheless, finding such problems raises students' awareness of this aspect of physics.

## 3.4 The problem of eliciting students' questions

In reality, some students may be out of the habit of asking open-ended questions, or have an educational background where it was not customary, or they simply don't want to risk looking foolish. Instead, they have learned to live with partial understandings and formula-following. This is a pity as question-asking is considered to draw students into the 'fundamental inquiry nature of physics' (Harper *et al* 2003). Perhaps it goes without saying that it can lead to better learning and may also inform the tutor of weaknesses in teaching and students' misconceptions. Harper *et al* (2003) had students keep a weekly, digital learning 'journal' in which they could ask two questions relating to their grasp of the subject, and one about what they would ask students if they were the tutor. The responses were useful in that they revealed conceptual difficulties (such as the difference between momentum and kinetic energy) and were used to inform subsequent teaching. Aflalo's study was a variation in which students work in groups to construct their questions and upload their approved questions to a course website (Aflalo 2018). These were answered by students in other groups, the process of question generation and answering taking

about an hour. Although neither studies requested explanatory, causal, 'Why?' questions, some were expressed, so an approach of this nature which specifically includes these could be useful. Variations of the approach may be readily constructed so that it includes and encourages causal thinking, problem noticing and finding, and, at the same time, gathers useful information about students' learning. For example, drawing on and supplementing these approaches, the questions might be:
- What did I not understand about this topic?
- What more would I like to know about this topic?
- What 'Why?' question which might lead to a practical investigation could I ask about this topic? OR What problem did you notice in this topic which could be investigated?

The nature of the questions and the constraints on them would, of course, vary with context. For instance, in a study by Etkina (2017), focusing on laboratory work, the students were asked: 'What other question could you investigate using the available equipment from this lab?' How responses are to be processed and feedback given may need thought, particularly if the class is large. An economical way of using 'Why?' responses is to select two or three likely to be of interest to many of the students and present them as a task. As an assignment, the students select one, explore and analyse it, reformulate it if necessary, and design an investigation to answer the question. The outcomes may be submitted in a variety of ways, ranging from e-posters to podcasts.

There is not always the time or equipment to do the proposed experiments this might generate, and matters of safety have to be considered. In practice, it may remain a paper exercise, but it practises students' creative thinking from the earliest step in the process and shows the origins of such thinking to be in asking 'Why?', problem noticing, finding, or posing, sometimes by the physicist who finds an answer, but also by someone who cannot do so and presents it to others for them to solve.

Students may, of course, be asked to research the topic in hand, and find unsolved problems in the literature. Being unsolved, such problems are likely to be demanding and beyond the resources of the teaching laboratory, although students may still speculate on them. Posed problems, constructed by the tutor, can be tailored to resources and the course. Not being their own questions, students may see them only as exercises of no great interest, but providing a choice of problems may overcome that. This posed problem approach is frequently the basis of inquiry-based learning (IBL), an umbrella term for a variety of approaches which leave the problem's solution for the student to construct with various degrees of guidance (Aditomo *et al* 2013, Lazonder and Harmsen 2016). IBL, however, can be a slow process, so a blend of teaching physics both as a body of knowledge and as an area of inquiry is likely to be more efficient, and also gives the students something to underpin the occasions when an enquiring mind is expected.

Where students have lost the inclination to ask questions, teaching which presents physics as investigation and inquiry some of the time can establish an atmosphere in which question-asking is normal and expected. This approach requires that physics is not always presented as a settled, complete, and entirely understood subject, but as something to explore, understand, and even play with (see also chapter 2). Students have the opportunity to become at ease, constructing problems, questions, and offering puzzles when they clearly see it as a part of the teaching and learning process. Nevertheless, they may find it surprising, even strange, to begin with. Problem noticing—'noticing a gap in understanding, [and] identifying and articulating its precise nature' is only a first step in the creative process and, at times, may be the easiest step (Phillips *et al* 2017). Whether the problem is noticed, found, or posed, it often has to be dismembered, refined, and re-formulated before it can be taken further.

## 3.5 Fostering students' thinking about problems

First, Phillips *et al* (2017) suggest that some students may need help in managing the state of confusion an unclear problem can produce. They may also need to be restrained from moving too hastily to an attempt to solve it before some clarification has been achieved. In particular, students may usefully be encouraged to see their confusion as 'intellectually enticing' and indicating a puzzle to solve.

Next, an early task for the student is to explore and analyse the problem. The aim is to draw attention to where further knowledge would support understanding and bring that knowledge to the problem. This is likely to reveal something of a problem's complexity and point to the need to treat it as a set of sub-problems, each to be explored and supplemented with additional knowledge as needed.

These sub-problems may be given an order of priority in the form of goals, and a plan of action constructed to achieve them. There is no fixed procedure for this but, through direct experience, perhaps with some support, the students have an opportunity to develop useful attitudes, skills, and general strategies (also known as heuristics). Algorithms and mechanical procedures are relatively easy to teach and use but there are no certain routines that provide explanations, answer questions, and solve problems. The result is that students may acquire strategies themselves through experience alone or not at all. In addition, some students do not seem to make a distinction between well-defined problems that lend themselves to routine solutions, and ill-defined ones that do not. We may not be able to give them a formulaic approach, but we do not have to leave it to chance alone. We can, at least, make students aware of different kinds of problems and of some strategies which may open ways to a solution, and we can demonstrate or model the approach for them. Heuristics offer a means of thoughtfully processing the problem to reveal clues to how it might be solved.

In mathematics, George Pólya asked, 'how is it possible to invent a solution?' He suggested that it helps to take four steps:
- Understand the problem;
- Plan an approach;
- Try it; and
- Review the outcome.

As a heuristic, this alone is not as helpful as it sounds. For instance, it leaves unsaid how to find an approach (see, for instance, a review by Voskoglou 2011). Since Pólya's early work, however, these steps have been found to be supported by various strategies. For instance, having students work in pairs has been found useful. One has the role of 'solver' while the other (who asks for reasons and clarification of thinking) monitors progress (Bilgin 2006).

### 3.5.1 A problem analysis heuristic

Some useful heuristics we might illustrate for students include one of *problem analysis* to reveal sub-problems. The process is commonly begun by producing an enriched means-end analytical map (digital or otherwise, individually or as a team using a shared surface). Each sub-problem may then be analysed further, or have attached to it what is known, and what needs to be known. What may supply the unknown may then be identified. This may allow some progress towards a solution. There are a lot of 'might' and 'may' in this because the outcome is not certain.

### 3.5.2 A qualitative account heuristic

Another heuristic is to construct a *qualitative account* of the problem in the student's own words (Lorenzo 2005). This puts aside formulae and mathematics, things which attract some students enormously, and, instead, describes the nature of the problem in terms of, for example, kinds of energy, the interaction of forces, linear and rotational. If this reveals a possible way forward, that is the point at which mathematics can be relevant.

### 3.5.3 Working backwards

If a possible path forward is still not evident, either for the problem as a whole or for one or more of its sub-problems, another heuristic is to consider what the solution might look like and *work backwards* from it. The aim is to reveal what will be needed and where it might come from. Depicted on an analytical map with a gap between the problem and the 'solution', connections may be made between the two. Both words and pictorial representations may appear in this approach as pictorial representations may suggest something not shown in the words (Maries and Singh 2018).

### 3.5.4 Using an analogy

An additional approach is to think in terms of: 'It's something like ...'. *Analogies* can point the way to productive thought and have long been used by scientists to represent and shape their initial notions. Of course, few processes are identical so analogies have limits and care needs to be taken when using them but, when other approaches stall, analogical thinking may help. For instance, it could be tempting to use Ayrton's sand ripples as an analogy for the ripples found in interstellar space.

Even though it is likely to fall short, the contrast alone may prompt ideas. Such an analogy may be depicted in a problem/solution chart to facilitate the comparison.

## 3.6 The tutor's contributions

Heuristics, however, do not come with a guarantee and other kinds of support can help.

For example, 'status overviews' in which students describe and reflect on their progress can help them see where they are and find a way forward more clearly. Ideally, students should eventually learn to do such overviews or reviews themselves. 'Prompts' which offer suggestions for ways forward can also be effective. 'Scaffolding' students' thinking through very demanding parts can similarly be helpful (Lazonder and Harmsen 2016). Sometimes, it helps to set out such guidance as a sequence. Of particular interest are questions, puzzles, and problems in need of explanations, from which predictions might follow. Students might usefully follow in the steps of the professional and develop such as explanation, perhaps by:
1. Making a clear, simple statement of the question.
2. Making sure of an understanding of the question/problem, and, when needed, collecting further information to support that.
3. Using diagrams as an aid to thought, placed in a temporal sequence, if appropriate.
4. Examining the diagrams analytically:
    a. What exactly needs explaining?
    b. Compared with where the phenomenon does/did not occur, what is different?
5. Asking what plausible causes there are, and choosing what seems to be the most plausible cause.
6. Using 'time-lapse' diagrams to represent the cause in action.
7. In conjunction with 1 above, identifying key features/variables on these diagrams, and known relationships.
8. Trying to relate these key aspects mathematically.

In practice, such sequences are really loosely ordered *aide memoires* for what is an iterative process, and this should be made clear to the students.

Students may not be fully aware of the role of models in physics. A model represents an idea or explanation and guides thought and action. Some models support understanding, explain past events, allow prediction, and underpin simulations. There are many kinds of model ranging from tangible objects (such as scale models), analogies, idealised models, and mathematical models (such as some models of the atom and the cosmos), to computational models (which may direct simulations). Finding a solution to a problem can mean finding or constructing a model to represent the phenomenon and running the model to its conclusion or backwards to an earlier state. The production of models does not seem to have received a lot of attention in the teaching of physics yet the part they play is considerable and worth attention.

Finally, in all of this, the student must learn that errors, mistakes, and false starts do not represent failure, but are a part of the process. At times, proposed answers to questions may be accepted only to be found to be wrong later. Perhaps a classic example is the collapse of bridges said to be caused by resonance as soldiers marched in step over them. After that, soldiers were required to break step on bridges, but closer investigation showed that they could not have imparted enough energy to bring the bridge down. Instead, faults in the bridge suspension were the chief cause (Donald 2017). Some would add that we should, at least, reflect on why an idea didn't work and learn from it.

## 3.7 There is no end to questions

Sometimes, students are slow to think of questions (or, perhaps, slow to express them in case others think them stupid). Providing a few examples can set the ball rolling. These might come from your experience, your colleague's ideas, or science publications such as *New Scientist*. From time to time, some publish collections of questions (and answers). Some are simple enough to induce students to talk without hesitation. (For example: Why, when sitting on a swing, does moving the legs to and fro make it move with increasing amplitude? (O'Hara 2014).)

The focus here has been on generating a question or a problem for students to think about as the start of the creative process in physics. But something that undergraduates may not appreciate is that successfully answering one question often leads to others. This message can be overlooked in the students' packaging of their answers. An explicit question asking, 'What other questions could this lead to?' could be a useful, concluding rider to their task.

## 3.8 Something to reflect on

Reflect on physics problems you have noticed in your own work.
- What were their origins? Where did they come from?
- When solving such problems, how do you go about it? What approach do you find useful?
- How would you describe your problem-solving approach(es) to undergraduates?
- What examples of problems embedded in the courses you teach could you use as examples?

There are also several sites with lists of unsolved, difficult problems in physics. While your students may not solve them, they may serve to illustrate that physics is an active and dynamic endeavour (see e.g. www.livescience.com and http://www.diva-portal.org).

And for your students:

### 'Beer mats make bad frisbees'

When asked to think of physics problems, students may believe that the expectation is too great: How can they think of problems worthy of a physicist? But this is to set the bar too high. It is not a matter of finding the difficult problem, or the one which moves a significant area of physics forward, but one which is from the world around them, interests them, and where the solution is not in a textbook or on the Internet. We might make the point by referring them to Cronquist's broken-off pencil points problem (chapter 2), or to 'Beer mats make bad frisbees', an investigation by Ostmeyer *et al* (2021) into why spinning beer mats (presumably to their disappointment) failed to perform as expected and ended up flipping backwards in flight. Our aim is to have students *notice* puzzles in their world and approach them with curiosity. Try setting them this *Ask why* task:

**Ask why:** As you go about your daily routine, ask 'Why is it like that? Why does it do that?' about what you see in the physical world around you. Bring an example that doesn't seem to have a ready answer to the next session to add to a collection of everyday physics problems which could be open to further investigation.

To help students understand what is mean by a heuristic, and to help them build their own heuristic repertoire to support their problem solving, try having them compile a chart of strategies (drawing on those illustrated in, for example, annotated Google Images or elsewhere). Their strategies could be organised to reflect the various stages commonly experienced in physics problem solving.

## References

Aatonen T *et al* 2022 High precision measurement of the W boson mass *Science* **376** 170–6
Aditomo A, Goodtear P, Bliuc A-M and Ellis R A 2013 Inquiry-based learning in higher education *Stud. High. Educ.* **38** 1239–58
Aflalo E 2018 Students generating questions as a way of learning *Act. Learn. High. Educ.* **22** 63–75
Ayrton H 1910 The origin and growth of ripple-mark *Proc. Roy. Soc.* A **84** 285–310
Bilgin J 2006 The effects of pair problem solving techniques *J. Sci. Educ.* **7** 101–6 https://files.eric.ed.gov/fulltext/ED495502.pdf (Accessed: 1 December 2022)
Donald D 2017 *When the Earth Was Flat* (London: Michael O'Mara)
Etkina E 2017 Using physics to help students develop scientific habits of mind *Sci. Educ.* **8** 6–21
Harper K A, Etkina E and Lin Y 2003 Encouraging and analyzing student questions in a large physics course *J. Res. Sci. Teach.* **40** 776–91
Irving Z C 2016 Mind wandering in unguided attention *Philos. Stud.* **173** 547–71
Kim Y, Seo H-A and Park J 2012 An analysis of the problem-finding patterns of well-known creative scientists *J. Korean Sci. Educ.* **33** 1285–99
Lazonder A W and Harmsen K 2016 Meta-analysis of inquiry-based learning *Rev. Educ. Res.* **86** 681–718
Lorenzo M 2005 The development of, implementation, and evaluation of a problem solving heuristic *Int. J. Sci. Math. Educ.* **3** 33–58
Maries A and Singh C 2018 Do students benefit from drawing productive diagrams themselves while solving introductory physics problems *Eur. J. Phys.* **39** 015703
Martinez M E 1998 What is problem solving? *Phi Delta Kappan* **79** 605–9
O'Hara M 2014 *Question Everything* (London: Profile)

Ostmeyer J, Schürmann C and Urbach C 2021 Beer mats make bad frisbees *Eur. Phys. J. Plus* **136** 769
Phillips A M, Watkins J and Hammer D 2017 Problematizing as a scientific endeavour *Phys. Rev. Phys. Educ. Res.* **13** 1–13
Qi J-Z and Zhang X 2020 A new cosmological probe using super-massive black hole shadows *Chin. Phys. C* **44** 055101
Savery J R 2006 Overview of problem based learning *Interdiscip. J. Probl.-based Learn.* **1** 9–20
Voskoglou M G 2011 Problem-solving from Pólya to nowadays *Progress in Education* ed R V Nata vol 22 (New York: Nova), pp 65–82
Williams M 2018 The missing curriculum in physics: problem solving education *Sci. Educ.* **27** 299–319

IOP Publishing

Creative Thinking in University Physics Education

Douglas P Newton, Sam Nolan and Simon Rees

# Chapter 4

# Creative thinking in practice: ideas

## 4.1 Introduction

Gleaming under harsh fluorescent lighting, an old red Routemaster London bus sits in the corner of an anonymous industrial warehouse as if ready to head out on the streets for its next shift. This bus, however, is not parked in a bus station but can be found at the CERN complex in Geneva, ready to take its passengers on a very different journey compared to touring around the streets of London. This unexpected addition to CERN is the 'ideas bus' (IdeaSquare 2022) and anyone who thinks of an idea is invited to come on board and ring the bell! The ideas bus is, quite literally, a vehicle to facilitate and highlight the central importance of ideas generation in physics and is the focus of this chapter.

In the previous chapter we explored the different ways to notice, find, and pose problems. We will now proceed to consider how to facilitate the development of ideas and tentative explanations. This is a challenging area for students and tutors as much as it is for researchers. To propose tentative (as yet unproven) explanations for phenomena requires self-confidence and a certain amount of courage. Karl Popper (1992) argued that science proceeded through bold ideas that were a departure from established scientific thinking, only some of which will prove useful. The scientific process necessitates challenging and questioning these explanations, but these explanations may also be dismissed or, worse, ridiculed by peers, colleagues, and wider society as illustrated by the vignette below.

## 4.2 Astronomer Copernicus

In one of Europe's oldest universities, the Jagiellonian University (Kraków, Poland), hangs an impressive painting by Jan Matejko (1838–93) of *Astronomer Copernicus*. In this painting Copernicus is depicted staring into the night sky, surrounded by instruments and his diagram of the heliocentric model of the Solar System. Open at his side is a copy of *De Revolutionibus Orbium Coelestium* which he published just before his death in 1543. Both the painter and astronomer are celebrated figures in their native Poland and are important symbols of national identity. Copernicus, of course, is recognized around the world for proposing the heliocentric model of the Solar System, a model that had previously been proposed by the ancient Greeks but was ignored in favour of the widely accepted geocentric model of the time. Furthermore, this 'Copernican revolution' is often credited with stimulating a wider creative scientific revolution questioning established theories and dogma. Living in a world now where the idea that the planets orbit the Sun is universally accepted and proven, it is difficult to appreciate living in a time with a very different understanding. However, gazing at the painting we can, perhaps, gain some insight and imagine joining Copernicus on the rooftops with the imposing cathedral looming in the background. Listening to his ideas, making observations, and taking measurements perhaps we would start to change our geocentric beliefs.

Copernicus's ideas can be traced back almost 30 years prior to the publication of *De Revolutionibus Orbium Coelestium* to 1514 and his theory was basically complete by 1532. However, Copernicus resisted calls to publish for fear of the scorn to which he would be exposed on account of the novelty and incomprehensibility of his theory. Perhaps this is who Paul McCartney was referring to when he wrote the Beatles song 'The Fool on the Hill' who 'sees the sun going down and the eyes in his head see the world spinning round' but 'nobody wants to know him'. In Matejko's painting we can get a sense of the challenges Copernicus faced, to not only imagine the world from a new perspective and collect evidence to support his theory, but also to have the courage and tenacity to persuade others. This is a situation we can see repeated throughout the history of science (the theories of evolution and continental drift for example), where new ideas from new perspectives challenge established beliefs. Just as this painting can transport us back 500 years, who will people be talking about in 500 years' time that fundamentally changed our understanding of the physical world? General and special relativity and quantum theory, for example, challenge our understanding of the world because they are beyond our observed perspective—just as we appear to be stationary on the surface of Earth as the Sun makes its regular journey across the sky. Therefore, the creative learner must not only be prepared to question established thinking, develop new ideas and gather evidence but must also be prepared to engage in debate and persuasion.

Jan Matejko's *Astronomer Copernicus* is one painting with many voices (National Gallery 2021) and illustrates how the creative teacher can use a resource such as this to explore the nature of physics and the stories that underpin the knowledge. Kant (see Buroker 2006) argued that we can never know the world as it is only the structures in our own minds which shape our understanding. The following activity (section 4.2.1) is designed to explore the structures in students' minds.

### 4.2.1 Student activity—structures in our mind

This activity is designed to explore individual mental representations and structures and is ideally carried out in small groups with one person reading out the instructions.
　i. Close your eyes and imagine an apple. What does it look like? What does it feel like? Can you smell it or taste it?
　ii. Open your eyes and describe what you experienced to someone else.

    iii. Close your eyes again and imagine that you are now shrinking in size such that you are now stood on the surface of the apple. You shrink even more and sink into the skin of the apple, passing between cellulose fibres. You become even smaller and are now amongst the chains of carbon atoms, then within an individual carbon atom and stood on the nucleus of the atom.
    iv. Keeping your eyes closed, look around and observe. What can you see and feel?
    v. Now you are starting to get bigger again and passing between the molecules and then on to the surface of the apple until you are back to your original size.
    vi. Open your eyes and describe what you experienced to someone else.

*Notes:* The apple is something everyone has experienced and is relatively easy to imagine; students may describe their favourite type of apple or picture a familiar apple tree. This initial part of the activity enables the students to engage and prepare for the second part. The view from the centre of an atom, however, is something nobody has seen or directly experienced but is a world that we frequently ask students to engage with. The second part of the activity enables students and tutors to explore this world through the structures in their own mind. Everybody's experience is unique and it is challenging to try and articulate this to someone else. As such, this activity promotes two aspects of creative thinking. It promotes idea generation by encouraging the student to imagine the sub-atomic world and it promotes elaboration by requiring the student to articulate this. In addition, justification can be brought into this exercise by asking the student to explain why they perceived the structure the way they did. What knowledge or evidence informed the visualisation? There are no wrong answers to this exercise as each person's imagined structure is their individual perception of reality. The exercise can help students develop more tangible associations with the sub-microscopic world that they will never directly experience.

## 4.3 Divergent thinking

Reflecting on any teamwork setting, we can all probably recall individuals who appear to be able and willing to contribute multiple ideas in response to a particular challenge or issue. These individuals are demonstrating the ability to engage in divergent thinking—the ability to generate multiple ideas (Guilford 1950). While some people may have a natural propensity to engage in divergent thinking, it is equally possible to facilitate opportunities to develop this capacity. Standardised tests with their emphasis on intensive practice to get 'the right answer' do not facilitate creative thinking and the development of new and unexpected ideas.

Key to fostering a positive divergent thinking environment is to ensure that critique is suspended and ideas are not discounted at this stage. Students may make suggestions where we know the science is wrong but as Taber (2016) pointed out, it can be all too easy to dismiss them and inadvertently stifle creative thought.

Guilford (1950) described divergent thinking as consisting of several components. These are:
- Fluency (generating many ideas).
- Originality (producing novel ideas).

- Flexibility (producing varied ideas).
- Elaboration (producing detailed ideas).

Creative thinking is domain and task specific. Diakidoy and Constantinou (2001) explored divergent thinking with a group university physics students in relation to fluency and originality of responses and type of task using open-ended and ill-defined tasks (section 4.3.1). The tasks either required an explanation of a phenomenon, a prediction based on a scenario or an application of a material. They demonstrated that the type of ill-defined task encountered affected the fluency and originality of the response with the greatest number of valid responses recorded for the application problem. The following activity (section 4.3.1) promotes divergent thinking with an application task.

### 4.3.1 Student activity—divergent thinking

Open-ended and ill-defined problems that allow multiple solutions are considered to facilitate creativity to a greater extent than well-defined tasks and problems (see Barron 1988, for example).

In this activity, students are presented with an open-ended question to promote divergent thinking. There should be no restriction placed on the suggestions made in terms of their feasibility or practicality. This is recommended to be a group activity.

  i. To establish the principles of divergent thinking, use an introductory task such as 'how many different uses can you think of for a paper clip?' (Guilford 1950).
  ii. *Fluency*—encourage the groups to generate as many ideas as possible without evaluating the ideas at this stage.
  iii. *Originality and flexibility*—these ideas can be reflected on in terms of how novel (originality) and varied (flexibility) they are.
  iv. Repeat the process but design the task within physics. Identify a problem that you would like the students to explore in your curriculum area. For example, Diakidoy and Constantinou (2001) ask their students to come up with as many ways as possible that a charged electrical piece of plastic could be used. The students can include additional properties as long as they are specified.
  v. *Elaboration*—the groups can be asked to provide more detailed explanations of their ideas. This provides the opportunity to explore the extent of their scientific understanding.

*Notes:* The paper clip example was used by Guilford (1950) for his original psychometric test. Open-ended and ill-defined tasks such as these can be challenging for students compared to tasks with definite and specific correct answers. However, this is the role of the creative teacher to facilitate these opportunities and promote creative learning.

## 4.4 Convergent thinking

Convergent thinking, in some ways, can be viewed as the opposite of divergent thinking. It is a tightening of the mind to reflect narrowly on an idea and is most associated with situations where questions only have one correct answer. However, convergent thinking has its part to play in creative thinking and may be thought of as involving critical thinking and the evaluation, selection, and refinement of ideas. Osborn (1957) proposed the Osborn–Parnes creative problem solving model where problem solving involves several steps. At the initial stage of the creative process, divergent thinking encourages all ideas to be recorded regardless of the appropriateness and practicality. Convergent thinking is then used to evaluate the suggestions and determine the best ideas to take forward. For example, rank the ideas generated in section 4.3.1 from best to worst. What criteria have you used to do this? How could the criteria be altered to obtain a different outcome? To develop as a creative thinker from the students' perspective, it is important to recognize when to use divergent or convergent thinking and how to best engage in this process. It is all too easy to engage in critical and evaluative thinking too soon, inhibiting the creative process. This may particularly be the case in group settings, where individuals may self-evaluate ideas and withhold suggestions because they are uncertain of their merit and wish to avoid criticism. The following student activity (section 4.4.1) is an example of combining divergent and convergent thinking.

### 4.4.1 Student activity—*The Martian*

This activity uses a fictional story to establish a narrative and the scenario. It combines divergent and convergent thinking within a problem task.

In the fiction novel *The Martian* by Andy Weir (2011) (also adapted for the cinema by Ridley Scott in 2015) Mark Watney is an astronaut stranded on Mars. In order to survive, Mark and the NASA team back on Earth must devise creative solutions to solve a vast array of problems such as energy generation, water production, and space travel. These problems relate to many different curriculum areas and can be used as a starting point for this activity, for example:

  i. *Divergent thinking*—present the students with the problem and ask them to come up with as many solutions as possible. Emphasise the importance of suspending critique and all possibilities should be included at this stage.
  ii. *Elaboration* can be useful at this stage to challenge the students to explain in more detail the science behind an idea.
  iii. *Convergent thinking*—now ask the students to evaluate the different ideas and narrow down to the best ideas. What criteria are being used to determine this?
  iv. The student teams present to the rest of the NASA team, the different solutions are evaluated, and a final vote is taken for the recommended solution.

*Notes:* It may take time to develop ideas, so it is recommended to provide the scenario and allow the students a significant amount of time to develop the full range of ideas. Examples from current research may also be used to illustrate to the students the creative thinking being applied to find solutions to these scenarios. Roberts *et al* (2021), for example, demonstrate how a composite of Mars dust, blood proteins, and urea can produce a construction material stronger than concrete.

## 4.5 Associative thinking

Ideas, of course, do not come from nowhere. The Ancient Greeks believed that ideas came from demi-gods known as the Muses who were the children of Mnemosyne (the goddess of memory). Using our memory can involve making new connections or associations between things we already know. Associative thinking is a key part to ideas generation as it is the process of linking one idea to another that may not have previously been connected. Students often see knowledge in isolation and struggle to make connections between different curriculum areas or across disciplines. However, successful development of associative thinking can be very powerful to promote creative thinking. Syed (2019) describes the importance of cognitive diversity within teams in order to promote new ideas. Groups are often dominated by people of a similar educational background and experience and, consequently, tend to arrive at similar outcomes. These outcomes may be very useful but may not be the most creative.

Reflecting on your own professional experience, have you worked as part of a team involved in curriculum design? Who was part of the team? Was it predominantly people from within your own department? How were alternative perspectives considered?

The following student activity (section 4.5.1) aims to help students develop their associative thinking skills and find new connections.

### 4.5.1 Student activity—new connections

This activity is designed to develop associative thinking skills and recognise new connections.

  i. Provide the students with an example of two apparently unconnected objects, e.g. a black hole and an onion, and challenge them to think of a connection between the two objects.
 ii. Now someone else suggests a new object and challenge the students to come up with a connection to this new object. The new connection cannot be the same as one previously used.
iii. The chain continues until a new connection cannot be made.
 iv. Now present the students with two curriculum areas and ask them to find connections between them such as astronomy and quantum physics.
  v. Now make connections with new curriculum areas such as electromagnetism.
 vi. Divergent thinking—the associations can also be assessed in terms of the quality of divergent thinking (section 4.3). To what extent are the ideas novel and varied?

*Note:* This type of task can be a good way to consolidate learning at the end of a topic.

## 4.6 Effective ideas generation

At this point, it is worth taking a moment to reflect on strategies to support ideas generation. The activities described so far typically require initially engaging in generating as many ideas as possible—an activity sometimes referred to as 'brainstorming'. Brainstorming has been shown to be an effective way of generating ideas, particularly when people are trained to use it (Parnes and Meadow 1959). There are several recommendations:

i. *Categories.*

It is helpful to initially identify categories and then use these as prompts for a series of brainstorming sessions. For example, in the activities above students may first be asked to identify several different categories that may be guided by curriculum themes and then to approach each of these categories in turn.

ii. *Free association.*

Key to successful brainstorming is to be non-critical, developing many ideas and letting one thought lead to the next in a relaxed and focused state of mind. This process of 'free association' can allow sub-conscious thoughts to appear (Firth 2019).

iii. *Keep on going.*

It is important to not get distracted from the process after generating one or two exciting ideas but to note these down and continue to try to think of more. The temptation is to focus on these one or two ideas as they look initially promising and miss out on other, potentially more creative possibilities. Also, do not stop and move on immediately if you feel that you are running out of ideas. The most creative ideas are often generated towards the end of a brainstorming session. It is not an inherently bad thing if you feel stuck or think that the ideas are not any good.

iv. *An appropriate length of time.*

It can take time for things to be remembered and for new associations to be formed. It is recommended that at least 20 min should be spent on a single brainstorming session (Firth 2019). Relevant ideas may also come to us at unexpected moments such as when we are out for a run or walk or in our sleep, so it is useful to be able to revisit an initial brainstorming session a few days later to explore further ideas.

v. *In groups or alone?* Brainstorming is stereotypically viewed as group activity with individuals huddled around a large piece of paper with 'post-it' notes in hand. However, group settings can harm ideas generation due to the pressures and anxiety of a group setting. Feelings of being judged can inhibit the flow ideas. It may be preferable for individual brainstorming to take place first and then to discuss the more promising ideas with trusted individuals, followed by more individual brainstorming.

## 4.7 Lateral thinking

Vertical thinking (characterised by logical and mathematical thinking) has been emphasized in education in various cultures and may be considered the dominant thinking skill in physics. Problems are solved through a series of 'vertical' steps that follow on from each other in a correct sequence (de Bono 2016). Students that are successful in physics develop this valuable thinking skill to a very high level. Lateral thinking, however, is taking a 'horizontal' approach to problem solving to identify new possibilities. In contrast to logical thinking, apparent wrong turns may ultimately be useful and lead to new perspectives. The mathematician Marcus Du Sautoy (2020) describes how he cannot rely solely on a simple step by step logical approach but this is combined with a more intuitive feel for other possibilities to be explored and not yet discovered. Such a capacity for lateral thinking is generally regarded as something purely within human cognition. However, the following vignette (section 4.7.1) describes one particular example where artificial intelligence has been able to demonstrate and promote lateral thinking and the development of new possibilities.

**4.7.1 AlphaGo**

The Chinese game of Go is the oldest game in the world still played and is regarded as embodying much of the creative and intuitive side of mathematics (Du Sautoy 2020). It is a game of territory where players take turns to place black and white counters on a 19 × 19 grid. If a player manages to surround a collection of their opponent's counters, then they capture those counters. The winner is the player who captures the most counters at the end of the game. The complexity of the game increases as more counters are added and it is estimated that a number with 300 digits would represent the number of games of Go that are possible (chess, in comparison, has around a 120 digit number). Consequently, it becomes impossible for a player to determine all the logical possibilities and consequences of a move and requires a more intuitive approach.

Hence, although the computer program Deep Blue demonstrated the capacity to beat the greatest chess players in 1997, it has long been considered that it would not be possible to design a computer program to win at Go. The astrophysicist Piet Hut said in the *New York Times* in 1997 'it may be a hundred years before a computer beats humans at Go—maybe even longer'. That day, however, came somewhat sooner than anyone expected.

In 2016 Lee Sedol, South Korea's 18-time world champion, was beaten 4 games to 1 by AlphaGo, the computer program developed by Demis Hassabis and the team at DeepMind. One of the most striking moments in this contest was move 37 in game 2. AlphaGo made a highly unconventional move that went against standard orthodoxy. Many experts were shocked by the move and considered it a mistake. To the contrary, however, as the game unfolded it proved to be a highly insightful move that, some fifty moves later, led to victory for AlphaGo. One expert observer commented, 'I've never seen a human play this move. So beautiful, beautiful, beautiful, beautiful.'

This move was both novel and useful and, therefore, demonstrates creative thinking. In so doing AlphaGo taught the world a new way to play and has led to new tactics. Hassabis described how the game of Go had got stuck on a local maximum. Established conventions meant that players had reached a peak in game play but AlphaGo had demonstrated that a new higher peak could be reached by breaking those conventions. AlphaGo is enabling humans to gain a deeper understanding of the game and revolutionized the way the game is played. Chinese Go champion Ke Jie stated 'Humanity has played Go for thousands of years, and yet, as

AI has shown us, we have not yet scratched the surface. The union of human and computer players will usher in a new era.'

### 4.7.2 Problem based learning

The mind handles information by forming representations that become established ever more rigidly over time (Newton 2000). Information that is used as part of one pattern cannot easily be used as part of a completely different pattern. Lateral thinking is a provocative attitude to challenge established patterns. With vertical thinking, an experiment is designed to show an effect whereas with lateral thinking an experiment is designed to develop ideas. In the classroom, therefore, it is important to try to encourage such contributions even if they may, in the first instance, appear to be inappropriate and a potential distraction.

The use of problem based learning (see also chapter 5) applied to finding solutions is one strategy to promote the development of lateral thinking skills. The problems should be ill-defined and students should not have all the required knowledge to develop appropriate solutions. An early stage in the task is for the students to identify potential areas that will need to be explored. Mustofa and Hidayah (2020), for example, investigated the effect of problem based learning tasks on student's lateral thinking skills in high school biology students. Lateral thinking skills were assessed in terms of four indicators: recognizing the dominant ideas of the problem, looking for different ways of looking at things, loosening rigid ways of thinking, and using random ideas to generate new ideas. In comparison to a control class not undertaking problem based learning tasks, the problem based learning class showed greater gains in all four of these areas.

However, developing teaching to promote lateral thinking can present challenges because teachers themselves have acquired their education mainly within the framework of vertical teaching. Vertical thinking can also be more appealing and relevant as it provides, from the start, the 'right' direction and an apparent effectiveness (de Bono 2016). In contrast, lateral thinking may be much more disruptive and lead to apparent 'failures' before a new idea and solution emerges.

## 4.8 Sticky creativity

Laurie Winkless is a physicist, science communicator and author of *Sticky: The Secret Science of Surfaces* and *Science and the City: The Mechanics behind the Metropolis* (Winkless 2022a). Laurie was a researcher at the National Physical Laboratory in London before joining the Nobel Foundation and developing a career in science communication. In this section, Laurie shares her experience of, and the importance of creative thinking in her career and physics more broadly.

*What does creativity mean to you?*
I personally need structure to be creative. The structure gives me a framework to work within and without. It provides a logical thread from which I can gather ideas, but then I can play outside those confines. It is not possible to have brilliant ideas and see them through all the time, but once I have an idea, I just try to remember that I am the

person creating that path, and no two paths are the same. That thought motivates me to develop ideas and try new things. But in truth, creativity looks different to everyone.

*How has creative thinking been important in your career?*
I grew up in a family strongly influenced by both science and performing arts. However, when I decided to pursue a future in science, I initially thought that to be a 'proper scientist', I would have to put my more artistic interests to one side. I was one of only a few female researchers in my group at the National Physical Laboratory (NPL). I sometimes felt that I didn't fit the stereotype image of a physicist and that this would be a barrier in my professional life. However, with the benefit of experience, I now recognise that the colleagues I really respected didn't have that same perception of me. Diversity in physics is key to developing a dynamic, creative workplace. I worked with many incredible people at NPL and loved my research. Leaving the lab was a giant leap into the unknown but I have always had a strong desire to learn new things and take on new challenges. I think it is important to be prepared to try at doing something and see how you get on, rather than not to try at all.

*How is creative thinking important in physics?*
The best physicists are the ones who can say, 'I don't know'. They are open minded to new ideas (divergent thinking) and do not immediately reject suggestions. However, they can filter ideas to focus on the most promising possibilities (convergent thinking). Experimental design is a creative process that may enable you to answer the question in more than one way and iteratively develop better experiments and find ever more interesting answers. Creative thinking is the ability to see the wider context and thinking more broadly about the implications and applications of the research. There are so many unanswered questions and creative thinking is essential to exploring these.

*How can creative thinking be developed in physics teaching?*
Physics is often taught as if the entirety of the field is 'a done deal'. The focus is on knowledge acquisition and applying equations, but this does not reflect the reality of working in science. Experiments in education settings are often activities to demonstrate an outcome that is already known. I think it is important to highlight that the process of doing science—something I learned about at NPL rather than school or university—is rather different. It is much more interesting to learn something yourself than be told something. We should remind students that there are still many unknowns and unanswered questions that will require the next generation of scientists to solve. This would have the potential to broaden the appeal of physics to more young people.

*How can teachers be supported to develop creative approaches to teaching?*
It is important to try and give teachers the time and space to think beyond the immediate curriculum and have opportunities to hear from current researchers and keep abreast of latest developments. This can reinvigorate a teacher's interest and provide opportunities to enhance their teaching.

### 4.8.1 The curling conundrum

Amongst the many interesting stories in *Sticky...* (Winkless 2022b), Laurie explores the curious behaviour of a curling stone. For many people, curling is something we may come across once every four years in the Winter Olympics with teams carefully sliding large granite stones down an ice rink on to a target area. It is interesting to observe the tactical and strategic play as well as the fine judgements and adjustments made to the movement of the stone on the ice—particularly characterised by sudden bouts of frantic sweeping with brooms in front of the moving stone. The base of the stone is concave rather than flat (like the base of a beer bottle) and this leads to some interesting and puzzling behaviour. For example, slide a beer bottle across a table and it will travel in a straight line. Apply some spin to the bottle and you should observe the bottle curl in the opposite direction to the spin. However, if spin is applied to a curling stone, then the stone curls in the same direction as the spin!

This curious observation can be used as the starting point of a teaching activity and ideas generation. Students can be asked to develop explanations for the curl of a beer bottle when spun (asymmetrical friction) and then challenged to suggest reasons for the opposite movement in a curling stone. Using divergent thinking, these ideas could be expanded, to think about interactions between different kinds of objects sliding on different surfaces. These ideas could be developed to design experiments to investigate these interactions such as a beer bottle on different surfaces or creating mini ice rinks with trays of water in a freezer. Thereby, the initial observation can be used as the basis of teaching sequence that can promote creative thinking as well as develop mathematical reasoning and experimental design.

The explanation for the observed direction of movement of the curling stones remains a source contention with two differing explanations suggested. Shegelski *et al* (1996) proposed the 'thin liquid film model' which proposes that increased pressure on the leading edge warms the ice, producing a thin film of water that reduces friction at the front of the stone. However, Nyberg *et al* (2013a) presented calculations for assumed distributions of the coefficient of friction in the contact between stone and ice that appear to indicate that the asymmetrical friction resulting from the thin liquid film model is insufficient to explain the observed motion of a real curling stone. As an alternative theory, Nyberg *et al* (2013b) proposed an explanation where the roughness of the stone's leading edge creates scratches that guide the direction of the curl (see Smarter Every Day 2022 for a summary of the ideas). The differing views of the two groups and the resultant debate is a good example of how explanations are developed and contested and the uncertainty of our scientific understanding.

## 4.9 Conclusion

In this chapter we have explored the components of ideas generation in physics and how creative thinking is fundamental to this process. Creative thinking itself involves divergent thinking to generate ideas and, within this, the approach can be considered in terms of fluency, originality, flexibility, and elaboration. Associative and lateral thinking should be applied to promote and develop novel ideas. Convergent thinking is then used to narrow down the ideas to those that are most promising and plausible.

## 4.10 Something to reflect on

Analysing your courses through the lens of creative thinking, can you identify parts that promote divergent, associative, lateral, and convergent thinking? Do module outlines and course information explicitly identify creative thinking as a course outcome? How could the course be designed to embed more opportunities to develop creative thinking and how can the importance of this be made apparent to the students?

## References

Barron F 1988 Putting creativity to work *The Nature of Creativity* ed R Sternberg (Cambridge: Cambridge University Press), pp 76–98
Buroker J V 2006 *Kant's 'Critique of Pure Reason': An Introduction* (Cambridge: Cambridge University Press)
de Bono E 2016 *Lateral Thinking. A Textbook of Creativity* (London: Penguin)
Diakidoy I A N and Constantinou C P 2001 Creativity in physics: response fluency and task specificity *Creat. Res. J.* **13** 401–10
Du Sautoy M 2020 *The creativity code: art and innovation in the age of AI* (Cambridge, MA: Harvard University Press)
Firth J 2019 *Creative Thinking: Practical strategies to boost ideas, productivity and flow* (Arboretum Books)
Guilford J P 1950 Creativity *Am. Psychol.* **5** 444–54
IdeaSquare 2022 The innovation space at CERN *CERN* https://ideasquare.cern/ (Accessed: 20 August 2022)
Mustofa R F and Hidayah Y R 2020 The effect of problem-based learning on lateral thinking skills *Int. J. Instr.* **13** 463–74
National Gallery 2021 One painting, many voices: Matejko's 'Copernicus' *The National Gallery* https://nationalgallery.org.uk/stories/one-painting-many-voices-copernicus (Accessed: 9 September 2021)
Newton D P 2000 *Teaching for Understanding* (London: Routledge)
Nyberg H, Hogmark S and Jacobson S 2013a Calculated trajectories of curling stones sliding under asymmetrical friction: validation of published models *Tribol. Lett.* **50** 379–85
Nyberg H, Alfredson S, Hogmark S and Jacobson S 2013b The asymmetrical friction mechanism that puts the curl in the curling stone *Wear* **301** 583–9
Osborn A F 1957 *Applied Imagination: Principles and Procedures of Creative Thinking* rev. edn (New York: Scribner's)
Parnes S J and Meadow A 1959 Effects of 'brainstorming' instructions on creative problem solving by trained and untrained subjects *J. Educ. Psychol.* **50** 171
Popper K R 1992 *The Logic of Scientific Discovery* (New York: Routledge)
Roberts A D, Whittall D R, Breitling R, Takano E, Blaker J J, Hay S and Scrutton N S 2021 Blood, sweat, and tears: extraterrestrial regolith biocomposites with *in vivo* binders *Mater. Today Bio.* **12** 100136

Shegelski M R, Niebergall R and Walton M A 1996 The motion of a curling rock *Can. J. Phys.* **74** 663–70

Smarter Every Day 2022 The controversial physics of curling *YouTube* www.youtube.com/watch?v=7CUojMQgDpM (Accessed: 10 March 2022)

Syed M 2019 *Rebel Ideas: The Power of Diverse Thinking* (London: Hachette)

Taber K S 2016 'Chemical reactions are like hell because…': asking gifted science learners to be creative in a curriculum context that encourages convergent thinking *Interplay of Creativity and Giftedness in Science* (Leiden: Brill Sense), pp 321–49

Weir A 2011 *The Martian* (New York: Ballantine)

Winkless L 2022a Physicist. Author. Science communicator *Laurie Winkless* https://lauriewinkless.com/ (Accessed: 15 March 2022)

Winkless L 2022b *Sticky: The Secret Science of Surfaces* (London: Bloomsbury)

IOP Publishing

Creative Thinking in University Physics Education

Douglas P Newton, Sam Nolan and Simon Rees

# Chapter 5

## Creative thinking in practice: experiments

### 5.1 Introduction

The challenges of engaging with science and developing understanding have long been recognised. As Claude Bernard (1865) said 'The science of life is a superb and dazzling lighted hall which may be reached only by passing through a long and ghastly kitchen'. If students are going to be make the journey to the dazzling lighted hall, the creative physics educator must use all the tools at their disposal to help them on their way. Fundamental to this is effective use of experimentation to illustrate the superb science and to illuminate the path through the kitchen to the dazzling hall beyond. It is through practical investigations that students can apply the scientific process and develop their conceptual understanding.

The Gatsby International Foundation report on good practice in practical science (Holman 2017) established ten benchmarks for good practical science. Key aspects of these benchmarks, in terms of promoting creative thinking, are that students should have the opportunity to:
- undertake independent research projects,
- undertake frequent and varied practical work, and
- be taught by expert teachers. In this context, this means teachers that have an expert understanding of creativity and how to develop pedagogies to promote it.

In this chapter we will explore creativity in experimentation and how to implement this in teaching and learning. We begin by considering the importance of the affective domain and gender equity in developing experimental pedagogy.

### 5.2 The affective domain

The affective domain focuses on student motivation, interest, and value (Wigfield and Cambria 2010). Affective outcomes of a learning activity can be just as important as developing scientific understanding but are often neglected in physics education (Ramma *et al* 2018). Bray and Williams (2020), for example, highlighted support, pressure, and emotional issues as the largest contributing factor to first year undergraduate students' perceptions of physics. Designing activities that promote positive feelings, attitudes, and engagement towards learning are likely to develop motivation in students and their willingness to develop independent learning skills. Borinca and Maliqi (2015) highlight the importance of the affective domain for creating an environment where there is motivation to learn and without which cognitive learning may be comparatively low.

The traditional teaching methodology consisting of a number of weekly lectures and a single laboratory session has been acknowledge as not the most effective method for students to learn (Newman *et al* 2015, Zawilinski *et al* 2016). Learning environments that focus on the process of science itself with cooperative learning environments that promote an inquiry based learning approach, including more student-centred and hand-on science activities, are recommended to improve learning outcomes (van Aalderen-Smeets and van der Molen 2015, Jeong *et al* 2021). This improvement in learning outcomes is due to the course fostering an environment that recognises and supports the interplay of the cognitive and affective domains.

## 5.3 Gender equity

Affective factors may be especially important for females with support from a parent, teacher, or close person being significant in determining course choices (Hazari *et al* 2007). The way science has developed, been undertaken, and taught is largely a product of male influence and involves practices that favour male success (Hazari *et al* 2007, Gilbert and Calvert 2003). The ASPIRES project (ASPIRES 2022) is a longitudinal research project studying young people's science and career aspirations in England. Interviews with students participating in the study overwhelmingly identified physics as 'masculine' and 'hard' (Macloed 2016). In the USA, fewer than 25% of undergraduate degrees awarded in physics were for females (APS Physics 2022) compared to just below 50% for chemistry. Skibba (2019) reports that women occupy less than 20% of postgraduate positions and struggle with obstacles and burdens at all career stages (something commented on by Laurie Winkless, the case study in chapter 4). In the UK 23% of A-level physics entrants were female in 2019 (compared to 39% for mathematics) (Keenan and Gupta 2022). The most substantial drop in female representation occurs between high school and university and choices of curricula content and activities may not be conducive to gender equity. This situation is long standing and systemic within the discipline. Stereotypes persist (either intentionally or unintentionally) that boys are better suited to physics (IOP 2020). For example, the popular television show the *Big Bang Theory* features four male physicists and engineers and their ditsy female neighbour. Therefore, to effect change it is essential to understand the factors that positively and negatively affect male and female engagement and performance in physics. If curricula are designed to support females to feel confident and be successful then more females will be retained within the discipline and further promote diversification within the field. The Institute of Physics report Improving Gender Balance (IOP 2017) highlighted several approaches such as ensuring a member of the senior leadership team develops strategy and leads change and providing training on inclusive teaching practices. Key to promoting creativity in any discipline or workplace is to develop diverse teams (Syed 2019) leading to new perspectives. Establishing greater gender parity is an important part of this. Furthermore, the pursuit of physics is a highly profitable enterprise for society and every member of society should feel equally able to engage in the discipline.

An important aspect to consider in relation to this is experimentation and within traditional physics pedagogy there are several examples of techniques that have been shown to be unsupportive or even detrimental to female attitude and performance. Significant amongst these are situations when male students dominate classroom discussions and activities (Stadler *et al* 2000) resulting in females engaging in fewer physics experiences (Bell 2001). This extends to cooperative learning situations such as in the physics laboratory with female students complaining of domineering partners, a lack of respect, and feeling that their partners understood much more than they do (Laws *et al* 1999). Bray and Williams (2020), for example, report comments from female first year undergraduate students such as 'we often found the boys' 'uncaring' state frustrating and insensitive'. In three-person groups comprising two male and one female there is evidence that the groups tend to be dominated by the males even when the female member was articulate and the highest ability member in the group (Heller and Hollabaugh 1992). Baker and Leary (2003) reported that girls wanted more interaction with peers, group work, and discussion. Therefore, it is important that group work and cooperative learning activities are well designed to promote appropriate interactions that benefit all members of student cohort. Haussler and Hoffmann (2002) highlighted the underrepresentation of content of interest to females in the curriculum and noted that content of interest to females was almost always interesting to males, but the reverse was not necessarily true. Therefore, educators are encouraged to embed gender inclusive strategies such as collaborative practical work and projects focused on everyday issues (Parker and Rennie 2002) and the application of concepts in broader perspectives such as applications to people or society and topics of interest to females. In addition, teacher behaviours should promote positive self-concept and confidence (Haussler and Hoffmann 2002).

## 5.4 Experimental demonstrations

A practical demonstration or experiment that challenges students' current understandings can generate puzzlement, curiosity, and a desire to explain the observations. This state of cognitive disequilibrium or 'variance' (Moon 2005) is a form of constructivism driven by the process of equilibrium and disequilibrium (Piaget 1971). This can also lead to challenging emotions such as feelings of discomfort, confusion, frustration, or even fear. It is important that students experience these feelings and recognise this as part of the creative process. The following vignette (section 5.4.1) illustrates how two great physicists mastered the art of experimental demonstrations to enthral audiences.

### 5.4.1 Faraday and Tyndall at the Royal Institution

Michael Faraday (1791–1867) received little formal education but as a young man he attended the public lectures of Sir Humphrey Davy at the Royal Institution in London. The demonstrations he observed sparked his curiosity and he made assiduous notes of the meetings. So impressed was Sir Humphrey Davy by the young man's interest that he offered him an apprenticeship in his laboratory and so began Michael Faraday's illustrious career in science.

Aside from his outstanding contributions to scientific research such as establishing the principles of electromagnetic induction, Faraday also personified the characteristics of an outstanding scientific communicator and the art of the experimental demonstrations to promote curiosity and creative thinking. For example, when discussing the everyday phenomenon of a burning candle Faraday (1865) states:

'…so wonderful are the varieties of outlet which it offers into the various departments of philosophy. There is not a law under which any part of this Universe is governed which does not come into play and is touched upon in these phenomena.'

He is explaining to the audience the connection between the everyday and apparently mundane phenomenon of a candle burning and the laws that govern the entire Universe. The importance of the burning candle is elevated from an everyday phenomenon to something of fundamental importance. He is promoting creative thinking, encouraging the audience to start to consider what they are observing from a new perspective. He also stimulates curiosity and hooks the audience to wanting to know more. Key to an effective narrative, in any context, is to create scenarios and questions that capture the audience's interest and its desire to find out what happens next. What has a candle got to do with the laws of the Universe?

Through simple experiments, he demonstrates how observations may challenge our understanding and lead us to ask questions. For example, he says:

'I will blow out one of these candles in such a way as not to disturb the air around it by the continuing action of my breath; and now, if I hold a lighted taper two or three inches from the wick, you will observe a train of fire going through the air till it reaches the candle.' (Faraday 1865)

A new result has happened. What is the cause? Why does it occur? How does the candle relight when the flame is not touching the wick? Faraday is not relying on a sophisticated or spectacular demonstration but rather the simplest of experiments.

John Tyndall (1820–93) followed in the tradition of Faraday and gave hundreds of public lectures at the Royal Institution as well as writing a dozen popular science books. These books contained many illustrations detailing practical demonstrations. He was highly regarded for his ability to communicate state of the art physics to a general audience and said of teaching:

'I do not know a higher, nobler, and more blessed calling.' (Lettice 2011)

A highly accomplished mountaineer, he used a climbing analogy in his concluding remarks of *Forms of Water* (Tyndall 1874): 'In the sweat of our brows we have often reached the heights where our work lay, but you have been steadfast and industrious throughout, using in all possible cases your own muscles instead of relying upon mine. Here and there I have stretched an arm and helped you to a ledge, but the work of climbing has been almost exclusively your own. It is thus that I should like to teach you all things; showing you the way to profitable exertion but leaving the exertion to you.' These words are as timely a reminder now, 150 years later, as they were then of the nature of scientific endeavour and learning.

## 5.5 Objects as analogies and metaphors

Objects need not necessarily be used only to explicitly explore a physics concept but they may also be used to represent concepts and to aid understanding. Many physics concepts require imagination as they are describing a world on a scale that is either too small (nuclear physics) or too large (cosmology) to be within the realm of normal human experience. Objects, therefore, can be a useful tool to support creative thinking in physics and develop understanding.

Gibbs (2011), for example, describes how a rubber sheet fixed to a bicycle wheel can be used to simulate gravitational fields. A set of weights hanging from the centre of the sheet enables the teacher to simulate the change in mass of the central object (planet or star). A ball bearing is then rolled on to the rubber sheet with the aim getting it to move at a tangent to a circle about the central mass to represent an orbiting satellite or planet. The depression of the sheet as a heavy ball bearing rolls across it stimulates the curvature of space formed by the gravitational field of a massive object.

Still (2017) uses Lego™ to explain particle physics with different colours and sizes of blocks representing different particles in the standard model. In this sense, the blocks represent the fundamental particles because they cannot be broken down any further but can be combined to make other particles. For example, a proton is made up of three bricks representing two up and one down quark.

In these contexts, these demonstrations are being used as analogies and metaphors to develop understanding of complex ideas when direct route to the new understanding is too difficult (Newton 2011). An analogue is used which functions similarly and shows a parallel effect to the targeted understanding (figure 5.1).

Activities 5.5.1 and 5.5.2 are designed to support students and lecturers to develop new experimental demonstrations or devise new analogies to explain concepts. Incorporating analogies into the curriculum can be a useful way to challenge the extent of student understanding as it requires the students to apply their knowledge in a new context rather than rote learning. Rhoad (2017), for example, incorporated analogies within a student assessment. The students were required to explain key terms within a topic with a definition, examples, and an analogy that would explain the meaning to someone without a scientific background. Development of the analogies proved to be the most challenging but also revealed where the students were more secure with their understanding.

It is important to be aware, of course, that analogies have their limitations (Duit *et al* 2001) and can be counterproductive if parts of the metaphor do not enhance

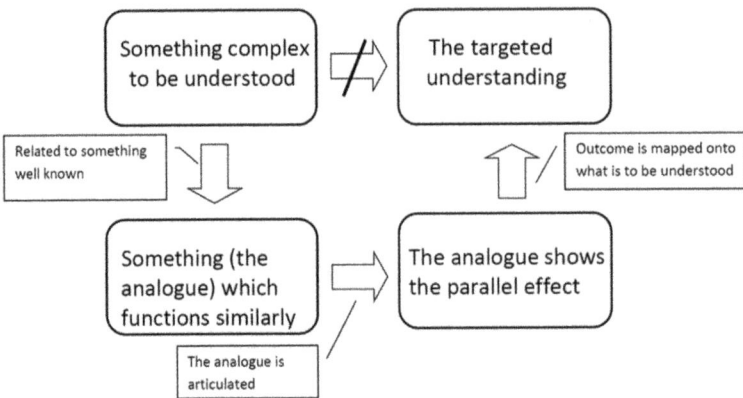

**Figure 5.1.** Using an analogy to enhance understanding. (Reproduced with permission from Newton (2011). Copyright 2011 Taylor and Francis.)

understanding (Beall 1999). If the metaphor is extended too far or taken too literally, misconceptions and incorrect mental models can develop (Heywood and Parker 1997). It is important to be aware if the analogy is culturally appropriate and relevant to all the students in order to support inclusive practice and gender equity. There may also be differences in the teacher's perspective and interpretation of the analogy compared to the students. However, this is also one of their values in promoting creative thinking. Once an analogy has been developed then critical thinking can be used to reflect on its limitations and potentially devise a better analogy. Metaphors are also useful from a social constructivist viewpoint (Ritchie 1999), promoting discussion about their strengths and weaknesses. Activity 5.5.1 suggests a framework for exploring metaphors with students and activity 5.5.2 adapts the activity for tutors.

### 5.5.1 Student activity—everyday objects

Everyday objects are excellent resource for promoting students' creative thinking. Students should be in small groups and appropriate ground rules established. This activity can be particularly useful either at the start of a course to explore current understanding or towards the end of a course to reinforce and explore links between different concepts.

**Stage 1**—Divergent thinking
Present the students with an object (e.g. a tennis ball) and challenge them to come up with as many different physics concepts as possible that can be illustrated with the object. There should be no limitations set on the ideas and no judgements made as to their validity.

**Stage 2**—Convergent thinking
The group selects the most promising idea and develops it into a demonstration with explanation for the rest of the class.

**Stage 3—Critical thinking**
The rest of the class provide feedback on the demonstration and reflect on limitations and improvements.

**Extension activity**
A variation on this activity is to make use of the physical environment rather than a specific object. Take the students on a walk around part of the campus or ask them to think of a journey that they regularly do. They are then required to try and link as many physics concepts as possible to different parts of the journey.

### 5.5.2 Tutor activity—challenging concepts

This activity is designed to develop new ideas for explaining challenging concepts. It is recommended to be undertaken in a small group with colleagues of different backgrounds and interests.

**Stage 1—Critical thinking**
Identify a section of your teaching that you or your students find challenging. Perhaps it is difficult to capture the students' interest, or it is difficult to understand. What do you think are some of the reasons for it being challenging?

**Stage 2—Divergent thinking**
Try to come up with as many examples as possible of different objects that could be linked to or help explain the concept. It can be helpful to do this in an environment with lots of different objects around (e.g. in a kitchen, museum, or sports shop) to promote ideas. Once again, no idea is too absurd and there should be no judgements at this stage.

## 5.6 Thought experiments

Whilst under house arrest in Florence and forbidden from publishing any books, Galileo completed his final book, *Discourses and Mathematical Demonstrations Relating to Two New Sciences*, in 1638. Despite the publishing ban the manuscript was smuggled to Holland and printed. In the *Discourses...* Galileo continues to demonstrate his unparalleled ability to engage in thought experiments to understand the physical world. He asks the reader to imagine dropping two objects of differing weights from a tower. According to Aristotle the heavier object will strike the ground first. But suppose that the two objects are connected by a piece of string. As the two objects fall, the heavier object should fall faster than the lighter object and the string would become taut. Does the lighter object act as a brake on the heavier object and slow its fall? Or is this now one combined object, with a greater weight than the original objects, that will strike the ground first? This contradiction led Galileo to conclude that all bodies must fall at the same rate, independent of their weight. Similarly, Einstein's capacity to engage in such imaginative feats led to the theory of special and general relativity. Kersting (2019) describes the use of an interactive warped-time model (www.viten.no/relativity) to help students explore the physics of free fall from both a classical and a relativistic perspective and engage with Einstein's thought experiments. Therefore, the creative physics teacher should exploit any opportunities to develop students' capacity to engage in thought experiments and hypothesise outcomes regardless of whether these can be practically demonstrated or not. For those interested in exploring thought experiments in more depth, Stuart *et al* (2018) provide an excellent insight into the history of thought experiments and their contemporary application.

## 5.7 Inquiry based learning

Formal science education is often criticised as it tends to emphasise content and facts rather than a method to discover new ideas and knowledge (Soares 2016). Traditional practical work requires students to follow a set of instructions leading to a predetermined outcome. These experiments enable students to develop practical and manual dexterity skills as well experience and exemplify knowledge. However, they have been criticised for their 'recipe' or expository approach with the focus on completing the required task rather than how it applies to the underlying concepts. The experiments confirm foregrounded knowledge and there is limited scope to promote creative thinking.

In response to these limitations, pedagogies have developed that reflect the scientific process and the nature of scientific inquiry. These less prescriptive and more open-ended approaches have been variously described as inquiry, discovery (or guided inquiry), process orientated guided inquiry learning (POGIL), problem based learning, and project based learning. All these approaches emphasise a student-centred approach, promoting student generated questions and the development of investigations to answer those questions.

Inquiry based learning is grounded in constructivism (Piaget 1971) and uses a learning cycle (Marek and Cavallo 1997) methodology of exploration (obtaining data), invention (deriving the underlying concept), and application of the concept to other situations. Inquiry based activities have an undetermined outcome and students generate ideas and procedures. Therefore, students are given more responsibility and there is a greater sense of ownership (Roth 2012). These activities promote higher order and creative thinking skills such as hypothesising, criticising, evaluating, analysing, and inventing (Raths *et al* 1986).

Inquiry based learning is more time consuming and demanding than expository approaches and places additional cognitive demands on students. They may be required to engage with new subject matter, laboratory equipment and techniques and problem-solving tasks simultaneously to complete the task (Linn 1977). Such an approach necessitates group work and the Social Pedagogic Research into Group-work (SPRiNG) project (https://www.spring-project.org.uk/) was a large scale project collaborating with teachers to facilitate group work. It was the largest quasi-experimental study in the UK to demonstrate consistent and significantly positive effects of group work on attainment, engagement, and attitudes. Whilst the focus was on the primary and secondary school phases, many of the findings are equally applicable to the higher education context and identified four key principles (see Fung *et al* 2018):

- Effective team dynamics in an environment that promoted mutual trust and respect.
- Classroom and grouping arrangements that are appropriate for the complexity and nature of the task.

- Appropriate task design.
- Supporting student autonomy.

It is essential for the creative physics teacher to develop a good understanding of the pedagogy of group work in order to deliver effective inquiry based opportunities such as project based learning.

### 5.7.1 Case study—experimental creativity

Shirish Pathare is Scientific Officer at the Homi Bhabha Centre for Science Education in India. He focuses on experimental physics and trains the Indian team for the International Physics Olympiad and trains teachers at secondary and undergraduate level. Shirish kindly agreed to meet up with us and share his thoughts on the importance of creative thinking for his career and physics more broadly.

*What does creativity mean to you?*
The starting point is being obsessed with a problem and spending some time with it. You need to be bothered by it and keen to solve the problem. You need to be prepared to try different possible solutions and open to things not working but evaluating and learning from these. Instinct is important for determining the most suitable way forward and even if the problem is difficult, you must not give up!

*How has creative thinking been important in your career?*
My focus is on experimental physics so the first stage where creativity is important is the experiment developmental phase. Every year I have to develop novel experiments for the Olympiad and I also develop low cost alternatives of undergraduate experiments such as a Michelson–Morley interferometer (Pathare and Kurmude 2016). The development of the low cost interferometer is an example of applying ideas from one context to another and the importance of instinct. In this instance, I had been working with speakers for some time and then realised that I could make use of a speaker to develop a novel design for the interferometer. I have also developed experiments to help students to understand difficult concepts such as the first law of thermodynamics (Pathare *et al* 2018). These experiments were designed to help students observe heat flow and explore the relationship with temperature change.

Second, there is the content delivery stage. The starting point for this is presenting ideas from the basis of what is already known and then to add in some surprising elements to generate curiosity and interest. The students are then encouraged to share their thoughts and ideas on the problem. The students then experiment with the set-up and try their ideas.

*How is creative thinking important in physics?*
Creative thinking is applicable to all disciplines, but the physics researcher must come up with alternative solutions and technological approaches to investigate physical phenomena. Similarly, the physics teacher must be very creative in coming with different ways in which explanations of physical phenomena can be explained to the students. This is particularly important when stereotypical views of physics can act as a barrier to engagement and students find the subject very difficult.

Creative teaching is also very important for classes where physics is not the main interest of the students but is a part of their degree programme.

*How can creative thinking be developed in physics teaching?*
The research shows us that active learning strategies help the development of conceptual understanding but to develop these strategies requires the teacher to be creative. The teacher should have a suite of strategies and can judge when and how to use these. To support teachers to develop creative thinking strategies it is important to recognise that teachers are also students and may possess similar alternative conceptions as their students. It is a major challenge to lead teachers to this sort of self-realisation but once it has been achieved, it is then possible to help the teachers appreciate the misconceptions that the students may also have and how to address them. Unless teachers undergo this process of overcoming these conceptual difficulties, they cannot appreciate the need to be creative in teaching and cannot be motivated or equipped to handle students' problems. It is then important to provide them with opportunities to learn about and experiment with alternative approaches and how these can be extended and developed for their context.

### 5.7.2 Project based learning

Project based learning (see also chapter 6) is an example of inquiry based learning that provides students with the opportunity to engage in real problem-solving and knowledge construction in authentic professional contexts leading to the development of authentic products (Brundiers and Wiek 2013). In Karl Popper's view, science starts with problems, rather than observations, and it is in the process of grappling with a problem that a scientist makes observations (Stanford 2021). Project based learning is an approach that has the potential to provide educational experiences that foster student autonomy (Martin *et al* 2017) and facilitates a learning environment aligned to real world situations (Balve and Albert 2015).

Krajcik and Shin (2014) identified six benchmarks for project-based learning:
- A driving question (this would be provided by a partner organisation).
- Learning goals (these can be a wide range of knowledge and skills outcomes).
- Educational activities (the pedagogical strategies employed).
- Collaboration among students (the projects necessitate effective collaboration to be successful).
- Use of scaffolding activities (how are the students supported appropriately towards achieving the learning goals?).
- Creation of tangible artifacts (e.g. a prototype project).

The solving of authentic problems and the production of tangible artefacts distinguishes this approach from other student-centred pedagogies. The creation of products is argued to be important because it helps learners to 'integrate and reconstruct their knowledge, discover and improve their professional skills, and increase their interest in the discipline and the ability to work with others' (Guo *et al* 2020). Whilst this approach has similarities to problem-based learning, Guo *et al* (2020) state that authentic problem-based learning emphasises knowledge construction whereas project-based learning focuses on knowledge application. A review of fourteen studies that adopted project-based learning in STEM subjects suggest that it increased the development of both learner's knowledge, collaboration, and negotiation skills (Ralph 2016).

The use of project-based learning provides excellent opportunities for incorporating authentic assessment methods assessing a diverse range of skills typically found in the world of work (Villarroel *et al* 2018). Balve and Albert (2015) describe an assessment model in a production engineering module that focuses on teamwork, presentation skills, written reports, and one-to-one discussions. Students may initially have difficulties adjusting to a project-based learning and the autonomy it provides. However, Balve and Albert (2015) report that within a few weeks most successfully adjust to the new mode of learning.

At Durham University third year engineering undergraduates engage in an industrial problem-solving module. Students work on a two-week live project with a local company applying their already acquired knowledge in manufacturing, mathematics, mechanics, electronics, electrical engineering, and IT in an industrial context. Project briefs can be viewed at Industrial Partnership Committee (2022). At the end of the two weeks the students present their findings via verbal and written reports. The approach has led to the development of novel and useful solutions in a wide variety of contexts.

Activity 5.7.3 is a simple framework to encourage thinking about developing a project-based learning activity. It may well be that practical constraints restrict your ability to apply a fully authentic project based learning model but are there aspects that you could incorporate to provide the students with a learning experience that incorporates some of these principles? For example, it may be that through engagement with local companies you become aware of challenges that you could use as case studies to contextualise learning.

### 5.7.3 Tutor activity—project based learning

Using the seven principles of project based learning (driving question, learning goals, educational activities, collaboration, scaffolding activities, and tangible artifacts), outline how a project based learning model could be applied to your teaching.

When approaching this, use divergent thinking and avoid allowing constraints to limit the flow of ideas. For example, at this stage you may have no idea what the driving questions will be but you may have some ideas of organisations that could be approached to be involved.

Once you have the outline, return to each principle in turn and brainstorm to generate lots of ideas for each of the stages. Then apply convergent thinking, to refine and select the most promising ideas to develop further.

### 5.7.4 Technicians

An essential resource to support the development of experimental curricula such as project-based learning is skilled laboratory technicians. Technicians can explore different methods, devise new pieces of equipment and ensure activities are organised as efficiently as possible. Engaging in development of new activities is an important part of technician's professional development and could contribute towards professional recognition such as through the Technician Commitment (Science Council 2022).

## 5.8 Something to reflect on

Incorporating project based learning into a curriculum can be a time consuming and labour intensive endeavour. Are there approaches in your institutional context that could help with this? For example, maybe there are funds available to support teaching innovations that could help? Such innovations may also strengthen the opportunities for increasing the student voice in curriculum design.

## References

APS 2022 Bachelor's degrees earned by women, by major *APS* www.aps.org/programs/education/statistics/womenmajors.cfm (Accessed: 15 March 2022)

ASPIRES 2022 ASPIRES research *UCL* www.ucl.ac.uk/ioe/departments-and-centres/departments/education-practice-and-society/aspires-research#:%7E:text=Longitudinal%20research%20project%20studying%20young,March%202017%20during%20ASPIRES%202 (Accessed: 17 March 2022)

Baker D and Leary R 2003 Letting girls speak out about science *J. Res. Sci. Teach.* **40** S176–200

Balve P and Albert M 2015 Project-based learning in production engineering at the Heilbronn Learning Factory *Procedia CIRP* **32** 104–8

Beall H 1999 The ubiquitous metaphors of chemistry teaching *J. Chem. Educ* **76** 366

Bell J 2001 Investigating gender differences in the science performance of 16-year-old pupils in the UK *Int. J. Sci. Educ.* **23** 469–86

Bernard C 1865 *Introduction à l'Étude de la Médecine Expérimentale* (Paris: Baillière)

Borinca I and Maliqi A 2015 The influence of teachers on increasing students' motivation to the Ismail Qemaili High School in the city of Kamenica, Kosovo *Psychology* **6** 915–21

Bray A and Williams J 2020 Why is physics hard? Unpacking students' perceptions of physics *J. Phys.: Conf. Ser.* **1512** 012002

Brundiers K and Wiek A 2013 Do we teach what we preach? An international comparison of problem-and project-based learning courses in sustainability *Sustainability* **5** 1725–46

Duit R, Roth W M, Komorek M and Wilbers J 2001 Fostering conceptual change by analogies—between Scylla and Charybdis *Learn. Instr.* **11** 283–303

Faraday M 1865 *A Course of Six Lectures on the Chemical History of a Candle: To Which Is Added, a Lecture on Platinum* (London: Griffin)

Fung D, Hung V and Lui W M 2018 Enhancing science learning through the introduction of effective group work in Hong Kong secondary classrooms *Int. J. Sci. Math. Educ.* **16** 1291–314

Gibbs K 2011 *The New Resourceful Physics Teacher* (School Physics)

Gilbert J and Calvert S 2003 Challenging accepted wisdom: looking at the gender and science education question through a different lens *Int. J. Sci. Educ.* **25** 861–78

Guo P, Saab N, Post L S and Admiraal W 2020 A review of project-based learning in higher education: student outcomes and measures *Int. J. Educ. Res.* **102** 101586

Haussler P and Hoffmann 2002 An intervention study to enhance girls' interest, self-concept and achievement in physics classes *J. Res. Sci. Teach.* **39** 870–88

Hazari Z, Tai R H and Sadler P M 2007 Gender differences in introductory university physics performance: the influence of high school physics preparation and affective factors *Sci. Educ.* **91** 847–76

Heller P and Hollabaugh M 1992 Teaching problem solving through cooperative grouping. Part 2: designing problems and structuring groups *Am. J. Phys.* **60** 637–44

Heywood D and Parker J 1997 Confronting the analogy: primary teachers exploring the usefulness of analogies in the teaching and learning of electricity *Int. J. Sci. Educ.* 19 869–85

Holman J 2017 Good practical science *Gatsby* www.gatsby.org.uk/education/programmes/support-for-practical-science-in-schools (Accessed: 12 March 2022)

Industrial Partnership Committee 2022 Welcome to the Engineering IPC *Durham University* https://eng-ipc.webspace.durham.ac.uk/ (Accessed: 20 March 2022)

IOP 2017 Improving gender balance *Report* Institute of Physics, London www.iop.org/about/publications/improving-gender-balance#gref

IOP 2020 Limit less—Support young people to change the world *Report* Institute of Physics, London https://iop.org/sites/default/files/2020-11/IOP-Limit-Less-report-2020-Nov.pdf

Jeong J S, González-Gómez D and Cañada-Cañada F 2021 How does a flipped classroom course affect the affective domain toward science course? *Interact. Learn. Environ.* **29** 707–19

Keenan O and Gupta J 2022 There are reasons girls don't study physics—and they don't include not liking maths *The Conversation* https://theconversation.com/there-are-reasons-girls-dont-study-physics-and-they-dont-include-not-liking-maths-182382 (Accessed: 9 May 2022)

Kersting M 2019 Free fall in curved spacetime—how to visualise gravity in general relativity *Phys. Educ.* **54** 035008

Krajcik J S and Shin N 2014 Project-based learning *The Cambridge Handbook of the Learning Sciences* 2nd edn ed R K Sawyer (Cambridge: Cambridge University Press)

Laws P, Rosborough P and Poodry F 1999 Women's responses to an activity-based introductory physics program *Am. J. Phys.* **67** S32–7

Lettice E 2011 John Tyndall—Science Communicator *Communicate Science* www.communicatescience.eu/2011/05/john-tyndall-science-communicator.html (Accessed: 9 May 2022)

Linn M C 1977 Scientific reasoning: influences on task performance and response categorization *Sci. Educ.* **61** 357–69

Macloed E 2016 What makes the girls taking Physics A level so exceptional? *UCL* https://blogs.ucl.ac.uk/aspires/2016/01/15/what-makes-the-girls-taking-physics-a-level-so-exceptional/ (Accessed: 9 May 2022)

Marek E A and Cavallo A M 1997 *The Learning Cycle: Elementary School Science and Beyond* (Portsmouth, NH: Heinemann)

Martín P, Potočnik K and Fras A B 2017 Determinants of students' innovation in higher education *Stud. High. Educ* **42** 1229–43

Moon J 2005 Progression in higher education: a study of learning as represented in level descriptors ed P Hartley, A Woods and M Pill *Enhancing Teaching in Higher Education* (London: Routledge Falmer)

Newman D L, Deyoe M M, Connor K A and Lamendola J M 2015 Flipping STEM learning: impact on students' process of learning and faculty instructional activities *Curriculum Design and Classroom Management: Concepts, Methodologies, Tools and Applications* ed DeMarco (Hershey, PA: IGI Global), pp 23–31

Newton D P 2011 *Teaching for Understanding: What It Is and How to Do It* (London: Routledge)

Parker L and Rennie L 2002 Teachers' implementation of gender-inclusive instructional strategies in single-sex and mixed-sex science classrooms *Int. J. Sci. Educ.* **24** 881–97

Pathare S and Kurmude V 2016 Low cost Michelson–Morley interferometer *Phys. Educ.* **51** 063001

Pathare S, Huli S, Ladage S and Pradhan H C 2018 Understanding first law of thermodynamics through activities *Phys. Educ.* **53** 025013

Piaget J 1971 *Biology and Knowledge* (Chicago, IL: University of Chicago Press)

Ralph R A 2016 Post secondary project-based learning in science, technology, engineering and mathematics *J. Technol. Sci. Educ.* **6** 26–35

Ramma Y, Bholoa A, Watts M and Nadal P S 2018 Teaching and learning physics using technology: making a case for the affective domain *Educ. Inq.* **9** 210–36

Raths L E, Wassermann S, Jonas A and Rothstein A 1986 *Teaching for Thinking: Theories, Strategies, and Activities for the Classroom* (New York: Teachers College Columbia University)

Rhoad J S 2017 Written assignments in organic chemistry: critical reading and creative writing *J. Chem. Educ.* **94** 267–70

Ritchie S M 1999 The craft of intervention: a personal practical theory for a teacher's within-group interactions *Sci. Educ.* **83** 213–31

Roth W M 2012 *Authentic School Science: Knowing and Learning in Open-Inquiry Science Laboratories* vol 1 (Berlin: Springer)

Science Council 2022 What is the Technician Commitment? *Science Council* https://sciencecouncil.org/employers/technician-commitment/?utm_source=google&utm_medium=cpc&utm_campaign=BG_ScienceCouncil&gclid=Cj0KCQjwnNyUBhCZARIsAI9AYlErDnx68o1kXQUrnDDmYCNh3AChuNtBANSfeRplxMKwxORWQevgz3saAo0zEALw_wcB (Accessed: 12 May 2022)

Skibba R 2019 Women in physics *Nat. Rev. Phys* **1** 298

Soares L 2016 Sciencing: creative, scientific learning in the constructivist classroom *Interplay of Creativity and Giftedness in Science* (Leiden: Brill Sense), pp 127–51

Stadler H, Duit R and Benke G 2000 Do boys and girls understand physics differently? *Phys. Educ.* **35** 417–422

Stanford 2021 Karl Popper *Stanford Encyclopedia of Philosophy* https://plato.stanford.edu/entries/popper/ (Accessed: 9 June 2022)

Still B 2017 *Particle Physics Brick by Brick* (London: Octopus)

Stuart M T, Fehige Y and Brown J R (ed) 2018 *The Routledge Companion to Thought Experiments* (London: Routledge)

Syed M 2019 *Rebel Ideas: The Power of Diverse Thinking* (London: Hachette)

Tyndall J 1874 *The Forms of Water in Clouds and Rivers, Ice and Glaciers* vol 1 (London: Appleton)

van Aalderen-Smeets S I and van der Molen J H W 2015 Improving primary teacher' attitudes toward science by attitude-focused professional development *J. Res. Sci. Teach.* **52** 710–34

Villarroel V, Bloxham S, Bruna D, Bruna C and Herrera-Seda C 2018 Authentic assessment: creating a blueprint for course design *Assess. Eval. High. Educ.* **43** 840–54

Wigfield A and Cambria J 2010 Students' achievement values, goal orientations, and interest: definitions, development, and relations to achievement outcomes *Dev. Rev.* **30** 1–35

Zawilinski L M, Richard K A and Henry L A 2016 Inverting instruction in literacy methods courses: making learning more active and personalized *J. Adolesc. Adult Lit.* **59** 695–708

IOP Publishing

Creative Thinking in University Physics Education

Douglas P Newton, Sam Nolan and Simon Rees

# Chapter 6

# Creative thinking in practice: applications

## 6.1 Introduction

Seeing how experimental physics can impact in the real world is critically important for university students. Not only does it show the impact of the work they're engaged in, but critically opens up avenues to employment beyond academia that they may have not previously been aware of.

One only has to look at destination data for physics graduates from the last few years to identify a large suite of industries where graduates end up being employed, including:

- Aerospace and defence.
- Education.
- Energy and renewable energy.
- Engineering.
- Health and medicine.
- Instrumentation.
- Manufacturing.
- Meteorology and climate change.
- Nanotechnology.
- Oil and gas.
- Science and telecommunications.

For doctoral graduates, work from the Southeast Physics Network of universities showed that 70% of physical sciences and engineering doctoral holders are working outside higher education (HE) three and a half years after graduating (SepNET 2016). Destinations were varied and included health, business, finance, consultancy, engineering, sales, and marketing in roles such as actuarial trainee, credit risk analyst, patent examiner, global energy manager, statistician, and aerodynamics engineer.

Critical to the undergraduate experience then is providing authentic opportunities which ensure employability skills development is inbuilt and that students' own creativity is exercised. In chapter 2 the application of creative thinking within physics and in the solution of practical problems more broadly was described. But such applications need a frame of mind which takes ideas in one or more mental boxes and brings them together in the problem to be solved. Students may have a tendency to keep their mental boxes firmly closed, so we may need to encourage a habit of thinking more widely. In this chapter we start by exploring a model for the creative process and then introduce two approaches which encourage students to creatively use their skills, namely:

- **Guesstimation** (a form of problem based learning) presents students with real-world situations and asks them to quickly guess a rough answer with very little information, e.g. how many pieces of popcorn fill a 100 seater cinema?
- **Team projects** (a form of project based learning), where students work in teams to solve a real-world problem set by an external partner (often a business) and develop a bespoke creative solution to the problem.

We'll conclude the chapter by considering what common underlying principles encourage student creativity in two case studies.

## 6.2 Frameworks for creativity in learning

Before we explore the world of applications of creative thinking in teaching and learning through some case study interviews, we will draw on a confluence of two frameworks, one from cognitive psychology and one from education, to help the construction of creative applications, namely combining:

- A framework for the creative process.
- A framework for teaching session design underpinned by the creative process.

### 6.2.1 A framework for the creative process

A seminal piece of work which first explored a framework for the creative process is that due to Wallas, first described in *The Art of Thought* (Wallas 1926). Despite being nearly one hundred years old, this framework is still frequently referred to by creativity researchers to this day (Cropley 2009, Ellwood *et al* 2009, Gallate *et al* 2012, Sadler-Smith 2015, Sio and Rudowicz 2007). Wallas' original framework (see also chapter 2) suggests four phases for the creative process, namely:

*Preparation*
This is the work of learning about a challenge and all the elements which feed into it and trying (and failing) to solve it using traditional methods in different ways. The first phase embeds the idea into the conscious mind to a point where it starts to leak into the unconscious.

*Incubation*
This is a phase where Wallas recommends disengaging with consciously thinking about a problem and distracting oneself through other kinds of activities far removed from the problem such as physical and social activity. This allows the unconscious mind to mull over a problem.

*Illumination*
Illumination is the process whereby the unconscious mind comes up with a unique solution to a problem and the conscious mind becomes aware of this. It's akin to the famous Eureka moment described by Archimedes. We've all experienced this to a lesser or greater extent when we become aware of a new way to think about a situation. This often leads to a positive frame of mind, and often the solution we come up with feels like it will almost certainly solve the problem.

*Verification*
Once we've got a creative solution through the illumination process, the new phase is to test it out—does it work? Often the answer is, surprisingly, yes—showing the power of the unconscious mind to aid us.

I'm sure if you're a physics tutor or researcher you'll be able to point to times when you've been through this process and recall it feeling exceptionally special. An immediate question is then, is it possible to design learning activities which can engage students in this process actively, as opposed to by chance?

In the next section we'll explore a framework for learning activity design which may help you try to model this framework in the classroom.

## 6.3 Designing a creative learning activity

We have developed a framework for a teaching activity which uses Wallas' four-stage model shown in figure 6.1. The preparation elements (diagonal stripes) provide

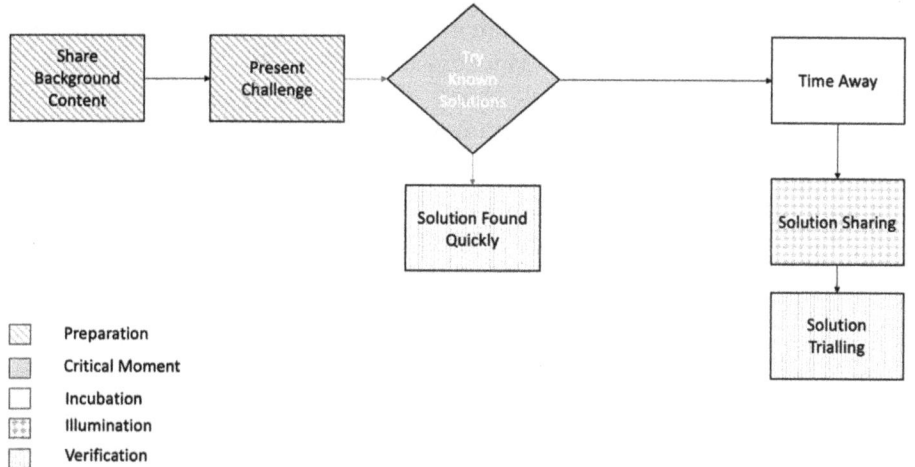

**Figure 6.1.** Designing in creative learning.

an opportunity for the teacher to share background information and present the challenge. Then comes a critical moment, students will either be able to solve the presented challenge easily and instantly move to a verification phase or not solve the challenge and move to the incubation phase. To design for creative approaches, the challenge which is addressed in the critical moment needs to be sufficiently robust and outwith the current understanding of the students to allow for incubation and illumination.

How can we design opportunities for practicing the application of creative thinking and how can we ensure that the same challenge will need a creative approach to solve it by all students? In the next section we explore two challenges as case studies that achieve this through two completely different pedagogical approaches and then we'll look to explore underlying commonalities between these two different approaches.

## 6.4 Case studies

Here are two case studies to illustrate how university physics tutors might encourage their students to apply creative thinking in physics related contexts.

### 6.4.1 Conceiving a guesstimation based curriculum (University of Glasgow)

Guesstimation is a form of problem based learning where students are presented with real-world situations and asked to guess a rough answer with very little information, e.g. how many pieces of popcorn fill a 100 seater cinema? A guesstimation survey for undergraduates was developed and evaluated by Macdonald *et al* (2013).

Dr Peter Sneddon, Senior Lecturer from the School of Physics and Astronomy at the University of Glasgow introduced the concept of guesstimation into the undergraduate curriculum several years ago. In the case study interview below Peter discusses the impact of this intervention.

*What was the approach?*
We developed the Glasgow GUESS, our adaptation of the General Undergraduate Estimation Skills Survey (GUESS) created by Macdonald, Burke, and Heiner at the University of British Columbia (Macdonald *et al* 2013). The assessment asks students both to complete a multiple choice set of questions and rank each answer based on their confidence and is used prior to the teaching and then repeated after to look for overall increases in performance and confidence.

*What issue were you trying to address when you developed this approach?*
Years ago we were trying to identify 'What did we all think were the important skills?' for physics undergraduates. One of the ones that came out of that was being able to make estimations to do order of magnitude calculations—basically, to be comfortable with being vague. It's one of the jokes I always do with first-year students—'I picked physics because there was always a definitive answer', and then as soon as you start doing experimental physics at university, you discover, there's never a definitive answer, and you've got to get used to that. We decided what we'd try and do is create a workshop for our first years where we would tackle skills in general, but a lot of it then ended up being built around the guesstimation type thing.

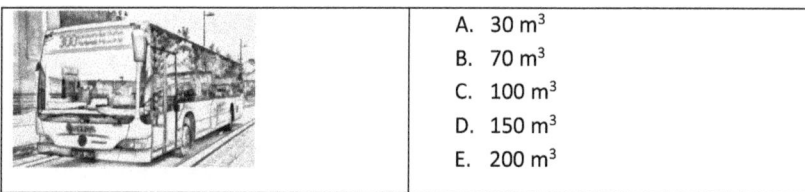

**Figure 6.2.** Multiple choice question: What is the volume of a standard city bus such as the one shown above?

*Give us an example of what's involved.*
If you present students with a question with exact figures and ask them to solve a problem using equations they are trained to be comfortable with this, but asking them to estimate say (as we do) the volume of a standard (single decker) city bus they find quite challenging. Using multiple choice questions though is itself quite useful as they can immediately remove clearly wrong answers. Take the bus example, adapted from the Glasgow GUESS questionnaire, in figure 6.2.

To arrive at an answer you think through the implications of a list of assumptions, e.g.:

- Bus is roughly a cuboid, with a square cross section.
- Height and width are both around 2.5 m—compare to size of person or
- width of road lane.
- Estimate the length around 10 m.
- Gives volume of ~2.5 m × 2.5 m × 10 m ~ 62.5 m$^3$ so B is the closest answer.

Students generally find this sort of question easier, whereas the following question (adapted from the Glasgow GUESS questionnaire) often differentiates between students of greater and lesser overall ability in terms of final score on the test.

What is the pressure, approximately, inside a typical party balloon?
  A. 0.7 atm
  B. 3.0 atm
  C. 1.0 atm
  D. 1.1 atm
  E. 10 atm

The key here is ruling out the impossible:

- Balloon would likely pop if B or E were the answer.
- If A was the answer, balloon would appear wrinkled.
- C and D the only reasonable answer; since it takes effort to inflate, likely pressure will be a bit greater than atmospheric, hence D.

*What makes this a creative approach?*
I really like trying to stop students putting facts in boxes and leaving them in boxes, which sometimes the curriculum can make them do. I think mixing it all together is incredibly important, and if they can get the hang of doing that then it makes their

lives a lot simpler. So, from that point of view, I think guesstimation does fit with that definition of creativity. It's about thinking about doing things different ways, pulling together all your knowledge and using that to help you.

*How did the students react?*
It's interesting, I think they got the idea. In the early years we did try and do a bit of work to see if it got better as the year unfolded. It was that great goal of all education research that's trying to show there's an actual impact from doing this. The end result—so one year we deployed this 'GUESS' at the start, and then we did it again at the end of the year, just to see. What we saw was their confidence improved a bit, but I wouldn't have stood up and said it was statistically significant, but it didn't get worse. In the workshops themselves we'd always get them to do it in groups and you'd wander around the room and you could see that some groups were really not comfortable and some were alright about it. But as the afternoon wore on, they got the idea. I think the more you went on about the fact, 'Look, it really doesn't matter if it's 5, or if it's 6. What matters is, is it 500, 600 or 5000, 6000?' You could see they were getting it. Comfort-wise, my impression was that most of them got more comfortable with it.

We would always get positive feedback. The students really enjoyed it, liked the way we did it, they found the level had suitable content. When we used it in the lab it certainly worked as an ice breaker. Because the students, we'd get them and they didn't know who anyone was but by the end of that day they were all nattering away at each other and coming up with half the questions.

*How have you adapted it over the years?*
What was quite good was it turned out, very rapidly, that it was relevant beyond physics because we got our university grant to explore the same skills idea across the college. So we went across all the physical sciences and, funnily enough, a lot of the skills are exactly the same, whether you're a chemist, a geographer, biologist, and out of that, we actually created an elective first-year course that has multiple sessions based around making estimations and approximations, and that's been a great success and the course has been full every year.

### 6.4.2 Team projects at Durham University

A second great example that encourages creative thinking in the real-world application of physics is in the use of project based learning (PBL) where students work with external bodies (companies, NGOs etc) to solve a problem set by the external body using physics approaches (see also chapter 5). Not only do the students get the chance to solve real-world problems, they also have to present to the external body in both a report and a team presentation. Many students involved in such programs go onto work with the external partner following their graduation.

Thomas Markham (2011) defines project based learning as follows: 'PBL integrates knowing and doing. Students learn knowledge and elements of the core curriculum, but also apply what they know to solve authentic problems and produce results that matter. PBL students take advantage of digital tools to produce high-quality, collaborative products. PBL refocuses education on the student, not the curriculum—a shift mandated by the global world, which rewards intangible assets such as drive, passion, *creativity*, empathy, and resiliency.' (Markham 2011)

An example of PBL is Durham University's Team Projects module which forms the basis for our second case study (Durham Physics Department 2022).

Durham University's Physics Department has been operating a project based learning module, known as Team Projects for over 25 years for its third year undergraduates. Team projects involve a group of up to six students working on a physics related problem set by either members of staff from the department or by local industry. The problems are as widely different as the companies that the students work with, but share some common elements:

- The problem will be 'real' in that there is no 'correct' solution and no script.
- It might, for example, involve building a piece of equipment, testing a product, designing a control system, etc.
- The module is assessed through a group at the end of the project, where external stakeholders are usually present, and through an individual short written report.

Professor Paula Chadwick Head of the Physics Department at Durham initially set up and managed these awards and received the Bragg Medal and Prize by the Institute of Physics in 2015 for her work in this area. In an excerpt from an interview, conducted for this book, Professor Chadwick talks about how this was set up, how important creativity is for physics students, and how this project allows students' creativity to flourish.

*What was the approach?*
Team projects involve a group of up to six students working on a physics related problem over a term. Experimental work is based in the university and the problem to be tackled is set by an industrial contact. Students will be expected to evolve their own approach to the problem, organise themselves, and work effectively as a team. Student performance is summatively assessed through a short written report on the project and an oral presentation. Practical sessions provide opportunities to obtain advice from staff members, for students to gauge their progress, and for staff to monitor progress throughout the duration of the module. In addition, a number of taught elements introduce ideas from project management by someone trained in project management.

*What issue were you trying to address when you developed this approach?*
The drivers were partly to alert physics students to what physicists could do in industry, because there isn't really a job labelled 'physicist' in industry as there is for say, chemists. So, that was part of the motivation. The other part of the motivation was getting students working together in teams. It's much more common now, but then it was quite rare. As we all know the reality of most working places is you're not working on your own, you're working in a group of people, and you've got to learn how to do it. So, that was another motivation. And the third motivation was to get

them to think a bit outside the box, to show them what they can do with the physics skills that they've got with a problem that's previously completely unknown to them. So, see where they can get from really a standing start in eight weeks.

*Give us an example of the sorts of industries involved?*
The National Health Service is becoming increasingly interested because they've realised there's a lot of physics behind their equipment. I've got a number of university spin-off companies too, including a bioscience company and other companies with broader physics principles they utilise, e.g. neutron tomography to find voids in infrastructure. Then there are firms like Carroll and Meynell who are transformer makers, Coltraco, a company utilising ultrasound, and EDF who run Hartlepool nuclear power station. We also do internal projects within the department to develop learning objects for the curriculum.

*How did the students react?*
I think initially they're very tentative, they're a bit scared by the whole prospect. Usually, about three weeks in, they're going, 'We have no idea how this is going to ever work.' By the end of the project, most teams say, 'That was great. Can we have another three weeks, please, because we really know what we'd like to do next?' Which I think is great. We always get very good feedback from the students.

*What would you consider to be a useful definition of creativity for a physicist?*
Students often think, 'Oh, well, if I want to be creative, I have to go and paint or go into theatre or whatever.' Science is also very creative. I think a lot of the creativity within physics is what you might call 'Only Connect', actually. It's about doing something in one area and connecting it with something else and being able to say, 'Hang on, this probably also applies to that, and I know how this works, even though I don't know how that works,' and trying it out. I think there's that. And I think there's also a bit of creativity in terms of building things experimentally, particularly when the students have a quasi-zero budget. We've done some great stuff with Raspberry Pis and Arduinos. All these things can run experiments, take measurements, and lots of students have learnt, 'We can programme them, we can sit them in the corner, we can get them taking measurements much faster than we can do it.'

*Thinking about the students, how important is it for you to try and encourage that way of thinking in the students?*
Yes. I think that is very important, actually. You stand with them and look at stuff, and they're looking at some graph or something popping out of the oscilloscope, and they go, 'I don't know what this is.' You go, 'Okay, let's think our way through it, shall we, and think what it could be.' And getting them to think about things in their everyday life that they do, and how that connects to the problem that they're facing —very often there is something there. Also, getting them to use the tools they already have at their fingertips to investigate things.

## 6.5 Discussion

Recalling figure 6.1 we should also ask: Do these case studies indicate that they have designed in the opportunity for creative thinking according to the model?

In the model we first introduced in figure 6.1 there are several key opportunities in the design we suggest would give opportunities to introduce creative learning among students, these are:

- *Preparation* (Share background content and present challenge)
    - In both case studies (guesstimation and project based learning) the students are presented with a challenge. The scale of the challenge clearly differs between the two case studies, with project based learning challenges far more likely to take significant time than guesstimation. However both require the sharing of background information and challenge, for guesstimation this may just be the question itself, but for project based learning is likely to be a more significant amount of information.
- *Critical moment* (Try known solutions)
    - In both examples the students will initially try to find a solution given their initial knowledge. With guesstimation this is likely to fail as the students will not have all the information they need to establish an exact solution. In project based learning the challenge will likely be too significant for known solutions to just work, after all its been identified by professional physicists working in industry as something that is worthy of further study.
- *Incubation* (Time spent away)
    - The time spent thinking about the problem may differ significantly between the two case studies. For guesstimation this will depend on how much experience with guesstimation and how comfortable they feel finding an approximate solution. For project based learning the time away may be significant and may not all be spent 'away' in the traditional sense. Students undertaking projects will learn new information and skills from undertaking the project itself and this may lead to many smaller Eureka moments along the way as they work towards a solution.
- *Illumination* (Solution sharing)
    - How and when students arrive at a solution itself will be specific to the student, the level of challenge they are under and how they manifest creative thinking. Both case studies certainly give the potential for this to occur, given the level of challenge and time afforded to the activity.
- *Verification* (Solution trialling)
    - If a Eureka moment occurs, which indeed it may, then both guesstimate and project based learning would afford the students an opportunity to try out their solutions. An assumption-based model derived in solution to a guesstimation activity could be tested against real-world knowledge, and a solution derived to a challenge presented in a project learning scenario could be tested in the lab or in the field depending on the nature of the project.

**Table 6.1.** Comparing 'normal' and 'authentic' assessment (after Wiggins 1998).

| Test based assessment | Authentic assessment |
|---|---|
| Requires the correct response | There isn't normally one correct response, rather students produce a high-quality product (talk, poster, artefact) |
| Must not be seen by students before they sit it | Should be known and understood by the students in advance |
| Is disconnected from real-world constraints | Is connect to real-world challenges, which students learn from undertaking |
| Contains questions which isolate particular skills or facts | Involves integrated tasks which involve a range of skills and knowledge |
| Includes easily scored items | Involves complex tasks, for which there is no one right answer and which are harder to score |
| Are one-off changes to evidence learning | Are iterative, students learn through 'doing' across the project |
| Provide a single score | Provides a range of evidence about a student's skills and knowledge from which a mark may be derived if required |

Each of the two case studies we have described differ from the 'traditional' assessments used when teaching higher education in physics, e.g. exams, lab reports, etc, in that they use authentic skills which are applicable in workplace settings. According to Wiggins (1998) an assessment is authentic if it:

- is realistic.
- requires judgement and innovation.
- asks the student to 'do' the subject.
- replicates or simulates the contexts in which adults are 'tested' in the workplace or in civic or personal life.
- assesses the student's ability to efficiently and effectively use a repertoire of knowledge and skills to negotiate a complex task.
- allows appropriate opportunities to rehearse, practice, consult resources, and get feedback on and refine performances and products.

Authentic assessments differ in several ways from traditional test based assessments as shown in table 6.1.

In table 6.2 we consider whether either of the two case studies could be classified as possessing an authentic assessment using the ideas from Wiggins (1998) discussed in table 6.1.

## 6.6 Conclusions

Beyond these examples themselves, what is apparent is an agreement between both academics we met in these case studies on what creativity in physics means. As Paula Chadwick puts it:

Table 6.2. 'Authentic' assessment (after Wiggins 1998) compared to both case studies.

| Authentic assessment | True for case study 1—Guesstimation | True for case study 2—Project based learning |
|---|---|---|
| Doesn't require correct response | There isn't strictly a single correct answer for guesstimation, a range of answers are allowed | There isn't a single correct response to a project |
| Might be known to students before they take it | As what's being tested is a student's ability to make an educated guess, seeing a question in advance wouldn't naturally help | Students may be aware of the project, but they learn through doing—undertaking the project |
| Is connected to real-world constraints | Although sometimes unusual, guesses are made about real-world scenarios | By being designed by employers this is implicitly built in |
| Pulls on a range of skills and facts | Guesstimation questions can pull ideas and concepts from across the curriculum and beyond. | Just like research, projects require a broad range of skills and knowledge |
| Isn't as easily scored as traditional questions | There isn't a single numeric answer and marking might be graded dependant on the quality of the guess | Projects reports are complex outputs which aren't scored like a traditional physics question |
| Involves iterative learning | Students learn how to make better guesses as they gain more experience of estimation questions | Is implicit as learning take places undertaking the project |
| Provides a range of evidence about a student's skills and knowledge | Developed to help students develop skills in estimation, but also pulls on knowledge of the broader concept | The project assessments if designed well will allow students to evidence the new skills and knowledge they've developed. |

'It's about doing something in one area and connecting it with something else and being able to say, 'Hang on, this probably also applies to that, and I know how this works, even though I don't know how that works,' and trying it out.'

And as Peter Sneddon confirms:

'I really like trying to stop students putting facts in boxes and leaving them in boxes, which sometimes the curriculum can make them do. I think mixing it all together is incredibly important, and if they can get the hang of doing that then it makes their lives a lot simpler.'

Both echo the sentiment that the structure of degree courses often segments and sorts physics knowledge into sub-disciplines, whereas creativity involves the removal of these barriers to allow students to look across their knowledge and experiences as a whole to address a particular problem in new, different, and innovative ways. Perhaps it's time for us to rethink our curricula in university physics and build a new way of thinking which removes many of these artificial barriers between sub-disciplines. Time will tell if this is possible, but the potential which could be unlocked by such a move seems very high.

Analysing briefly what the case studies reveal of the students' perceptions of these forms of teaching reveals an interesting challenge. In each example staff notice that students are somewhat anxious at first when dealing with new ways of working. This is natural, but the transition from anxiety to enjoyment seems quicker here and perhaps this is a hallmark of a creative approach to teaching in which even if the perceived stakes are high, students are given much greater autonomy than in more traditional forms of teaching. This then is the hallmark of the application space, allowing students to develop creative solutions to problems with an application beyond their course itself opens up a level of autonomy and embeds a key employability skill in the curriculum.

We also see that these case study university teachers have an informal understanding of a key aspect of creative thinking, namely that its essence is the bringing together of information and ideas to produce a solution (see Koestler's bisociation, chapter 2). This underpins the definition provided in chapter 1, that it is the production of something new or novel, at least to the student. Of course, this also has to be plausible or otherwise appropriate and, if the outcome is rewarding, it adds value. A shared understanding of what is meant by creative thinking in physics is essential if there is to be fruitful discussion, planning of activities, and assessment.

## 6.7 Something to reflect on

The practice of physics involves a variety of assessments, some which encourage creativity and link to real-world applications and some not.

- Readers might reflect on their own modules and assessments. Do these currently allow the creativity amongst your students you'd like to encourage or not? And, if not, what can you change?
- In practical terms, even if you're unable to change the summative assessments you use, how might a formative task be used to encourage greater creativity amongst your students?

## References

Cropley D H 2009 Fostering and measuring creativity and innovation: individuals, organisations and products *Proc. for the Conf. 'Can Creativity be Measured' (Brussels, May 2009)* 257–78

Durham Physics Department 2022 Durham Team Projects online *Durham University* https://durham.ac.uk/departments/academic/physics/labs/level-3/team-project/

Ellwood S, Pallier G, Snyder A and Gallate J 2009 The incubation effect: hatching a solution? *Creat. Res. J.* **21** 6–14

Gallate J, Wong C, Ellwood S, Roring R W and Snyder A 2012 Creative people use nonconscious processes to their advantage *Creat. Res. J.* **24** 146–51

Macdonald A J, Burke S A and Heiner C E 2013 Development of a General Undergraduate Estimation Skills Survey (GUESS) *Physics Education Research Conf. 2013 (Part of the PER Conference series) (Portland, OR, July 17-18, 2013)* pp 237–40

Markham T 2011 Project based learning—a bridge just far enough *Teach. Libr.* **39** 38

Sadler-Smith E 2015 Wallas' four-stage model of the creative process: more than meets the eye? *Creat. Res. J.* **27** 342–52

SepNET 2016 Where do physics doctoral graduates go? *SEPnet PGR Destination and Placement Report* South East Physics Network p 1–11 https://sepnet.ac.uk/wp-content/uploads/2016/09/Where-Do-Physics-Doctoral-Graduates-Go-SEPnet-PGR-Destination-and-Placement-Report-2016.pdf

Sio U N and Rudowicz E 2007 The role of an incubation period in creative problem solving *Creat. Res. J.* **19** 307–18

Wallas G 1926 *The Art of Thought: a Pioneering 1926 Model of the Four Stages of Creativity*

Wiggins G 1998 Research news and comment: an exchange of views on 'Semantics, psychometrics, and assessment reform: a close look at 'authentic'assessments' *Educ. Res.* **27** 20–2

**IOP** Publishing

Creative Thinking in University Physics Education

Douglas P Newton, Sam Nolan and Simon Rees

# Chapter 7

# Recognising creative thinking in physics

## 7.1 Uncertainty and assessing thinking competences

Words such as assessment can have connotations of precision and accuracy, but the assessment of students' competence, aptitude, or potential in any discipline is not the precise practice that some think it is. At times it may depend as much on extraneous factors as it does on the intellect—on a different day, there may be a different outcome. In short, a particular score may seem precise, but in reality it is only an indicator of a student's ability. In the absence of mind-reading, we give tests and set tasks to find evidence of what we cannot see. As with many such tests, for practical purposes there has to be a trade-off between how good the test is at indicating creative competence and the time it takes to administer the test and assess the answers.

This is certainly true of creative thinking—the kind of thought which can get it wrong yet still be worthy of credit and praise. In physics, we have the habit of setting students tasks and problems which have one answer and we know what it is. As Diakidoy and Constantinos (2001) argue, physics students are typically expected to describe, calculate, and do practical work, and these tasks 'have been carefully pre-specified as to their procedures and outcomes' in order to support the learning of concepts, theories, and procedures. This is not something inherently bad—there is much that can learnt in this way, and it makes assessing students' learning easy—we know the answer and the students' are either right or wrong.

The products of creative thinking are less predictable. Faced with an ill-defined problem, students may produce many plausible solutions, including some we did not expect and are not in our mark scheme. This means that creative ideas cannot usually be judged mechanically; we have to engage with them in a different way. Some have been tempted to see such ideas as beyond measurement, and others have said that making it a high-stakes test suppresses creative thinking as students avoid risk taking. But this is to put students' learning and the effects of a course out of our reach. It also makes formative feedback, at best, difficult, and so improvement is left to chance. How, then, can we assess, or, at least, *recognise* creative competence? Given a tutor's other duties, complex, time-consuming methods are impractical. They have two sources of information. The first is the *product* of creative thought.

## 7.2 Assessing the product of creative thought

The product could be, for instance, a problem (if noticed or found), an idea (tentative solution, explanation), a test of the idea (e.g. experiment design), or some application (to explain other phenomena, or to solve a practical problem of need). The questions are, 'How novel is it (at least, to the students)?' and 'How appropriate is it (given the context)?' If the answer to neither of these is zero (recalling that Creativity = Originality × Appropriateness), then we might add, 'How clever/ parsimonious/satisfying is it?' for bonus points. Bearing in mind the trade-off between 'precision' and available time and an appetite for tedium, one popular, economical, and useful approach can provide useful information and also inform feedback to students.

### 7.2.1 Assessment by consensus

Given that you are a physics specialist and experience has taught you what students are generally able to produce, one way of judging creative competence is to accept that you know it when you see it. The approach, devised some time ago by Amabile (1996), can be seen as assessment by consensus and can produce remarkably consistent rankings of students' performance. In essence, the students' work is ranked subjectively, from most creative to least creative, by, for example, two tutors. There can be a high degree of agreement between tutors, and where there is disagreement, it is usually resolved quickly by discussion. This is relatively easy when the number of students is small, but a diamond ranking organisation can reduce the labour when there are many students (figure 7.1 shows one for 27 students). If the task results in a product poster, the students may also learn by

A<br>
BBB<br>
CCCCCC<br>
DDDDDDD<br>
EEEEEE<br>
FFF<br>
G

**Figure 7.1.** Here, student responses are ranked from the one seen as the most creative (A) to the one seen as least creative (G). The assumption is that there is a roughly normal distribution with most students of average performance (D), while progressively fewer will appear above and below the average. In practice, adjustments to reflect obvious deviations can be made.

engaging in the ranking process themselves. While this refers to the ranking of a product for holistic evidence of its creativity, it might be used to rank novelty, plausibility, and, perhaps, parsimony separately.

The approach, however, puts students in order but does not assign a numerical score. One solution is to allocate score ranges to each row in figure 7.1, so that A could be assigned the band of 90%–100%, and so on. The actual score awarded in a given band depends on how the student's product is judged in comparison with others in that band.

### 7.2.2 Simple rating

Another approach is to rate the attributes of interest subjectively on a simple scale. Suppose, for instance, that your interest is in the students' *problem reformulation* and *consideration of plausibility*, and in the use of aids to thinking, such as *digital tools* and *collaboration*. Balchin (2006) used a grid for grading such features, as in table 7.1. In practice, the list can be longer. If criteria are constructed for each item this can have an appearance of objectivity. Nevertheless, the scores produced should not be treated as precise or equal interval measurements of such attributes. Again, ratings by more than one person may be compared, discussed, or averaged. Nevertheless, whatever approach is used, we should be aware that a student's performance on only one task may not, in itself, be a sufficient indicator of creative competence, but could offer useful feedback to students about where they might give more attention.

### 7.2.3 Putting numbers to it

Calling for more time and effort, there are more complex ways of gauging creative abilities. Some interested in seeing if a particular course can make students think more creatively have used pre- and post-tests of *idea generation*, *thinking contrary to current understandings*, and *concept combination* (e.g. Kohl *et al* 2011). For example, thinking of as many ways as possible to convert wind energy into other forms of energy has been used to prompt idea generation; ways of exceeding the speed of light would be contrary to current understanding. The number of responses to such questions, their detail, and how quickly someone can change from one kind of

Table 7.1. Each student's performance is scored for each item. In practice, other or additional items are likely, according to the teaching goals and the competences being assessed.

|  | Weak | Moderate | Strong |
|---|---|---|---|
| Problem reformulation |  |  |  |
| Plausibility or appropriateness |  |  |  |
| Use of digital tools |  |  |  |
| Collaboration |  |  |  |

response or approach to another, have been used as indicators of creative potential in physics.

Although this has a flavour of objectivity, its relevance can be questioned. There can also be a tendency to focus on an ability to generate many ideas, without regard for their quality. In a subject where plausibility is important, not just any wacky idea counts. Assessment has to include some aspect of idea quality, appropriateness, or fitness for purpose. Of the three aspects of creative thinking discussed at the outset, originality, plausibility/appropriateness, and impression/satisfaction, they may not be of equal importance or weight. A highly original but completely implausible idea, however elegant, may attract little favour. The importance given to each of these may vary from domain to domain but, in physics, figure 7.2 is likely to be a common perception of their relative importance.

Diakidoy and Constantinos (2001) point out that much that goes on in physics is directed at solving ill-defined problems and this is where creative thinking is needed. They focused on assessing the two dominant aspects of figure 7.2, originality or novelty of ideas, and their plausibility or appropriateness in this kind of problem. They constructed ill-defined problems which required their university students:

- to provide possible explanations of mechanisms behind a phenomenon, or
- to make predictions about a physical situation or event, or
- to think of possible ways of using a phenomenon, item, or device.

Asking such questions offers a way of integrating the expectations of creative thinking with an existing course and making it a part of learning physics, rather than an add-on. Their scoring of responses is instructive as it takes plausibility (appropriateness) into account. After sifting each student's responses to remove near duplicates of ideas, those left are judged for plausibility (appropriateness) to give the number of 'valid' responses made by each student. This removes responses

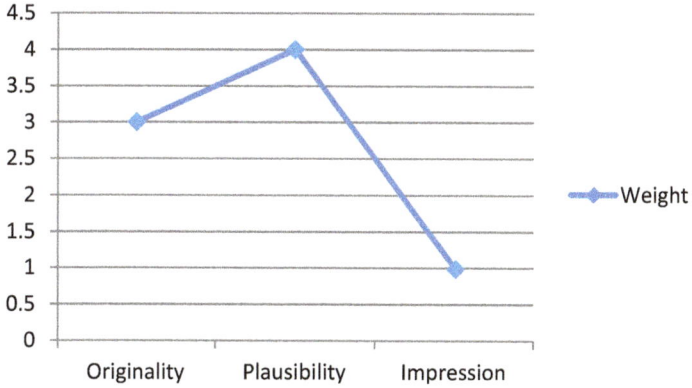

**Figure 7.2.** The importance (or weight on a scale of 0–5, *y*-axis) given to the aspects of creative thinking (*x*-axis) can vary from domain to domain. This is an illustration of their relative importance in some aspect of physics.

that are variants or are scientifically irrelevant and is done by direct inspection, something that Diakidoy and Constantinou say was relatively easy.

The frequency of occurrence of responses across the whole sample is then noted. Those given by fewer than 5% of students are given a score of 3, those by fewer than 15% of students are given a score of 2, those by fewer than 50% are scored as 1, and the remainder receive a score of 0. This rewards students for original, plausible thinking. (The percentages are not critical and simply sort responses by rarity.)

This approach provides a way of quantifying creative responses to ill-defined problems calling for creative solutions. Nevertheless, one ill-defined problem alone is unlikely to give a firm assessment of a student's competence or short-comings. A profile of student responses accumulated over time is likely to give a better view of these. Readers may recall that the third element in figure 7.2, impression/satisfaction, is seen as adding to the appeal of a creative idea, but is not an essential element of it. It does not seem to be something that is formally assessed, and would again be difficult to quantify, but may be recognised by tutors. Although this brings some objectivity to the assessment, it is at the cost of time and effort and physics tutors may prefer quicker approaches, such as those outlined above.

## 7.3 Assessing the process of creative thought

Often, assessment begins and ends with the product of creative thinking. How a student thinks a way through to a solution, the *process*, is generally less readily available. That thinking may have focused on problem sensing (e.g. raising a question or puzzle), task exploration and preparation (e.g. analysis, information collection), concentration (attempting to find a solution), incubation (e.g. off-task processing when concentration was unsuccessful), illumination (a potential solution appears or evolves), verification (an exploration of appropriateness), and persuasion (e.g. a gathering other evidence to support the solution). Unless a student talks about these at the time, awareness often fades and they can't be recalled. But if we could access the process, we might judge its strengths and weaknesses and offer relevant advice on what needs attention and how it might be improved. Again, we have to consider how long the assessment might take, and what staff demands it makes. The approaches below range from relatively undemanding to more time-consuming.

### 7.3.1 Just ask

*Difficult or easy?*
Probably the simplest approach is to ask students which parts they found to be difficult and which were easy, and what made them so. This might be done in a group tutorial and advice can be provided for all at the same time. If you feel a degree of anonymity would help, you might have the students vote electronically for each aspect of interest.

*One-to-one recall*
A second way is to have students explain how they arrived at their product during a tutorial when, for example, an assessment by consensus is presented. Unless this is very soon after the event, students are unlikely to recall their flow of thought for you.

Pointing to particular strengths and weaknesses of the product, however, can prompt them to reconstruct their thinking for you. Given that this takes place in a one-to-one tutorial, this inevitably takes time, although it seems to have some potential for providing insights into students' thinking.

### 7.3.2 In writing

*Written commentary*
Another approach is to ask students to present their creative work as it was done, warts and all, accompanied by a commentary or rubric in a margin which briefly indicates their thinking as the work progressed. Of course, this entails reading the students' rubrics and not simply looking at the final product.

*Self-reflection*
Students may be asked to think about their thinking and add a paragraph to their solution which evaluates it, describes the approach to the problem, identifies difficulties and obstacles, and states how they were overcome (if they were). It may help students to reflect in this way if you provide a short list of what interests you (e.g. problem exploration, understanding, reformulation, simplification, idea generation, plausibility consideration) and students rate the difficulty they had with them on a simple scale from 0 to 5, and indicate where they would welcome advice. Reflecting on and evaluating one's own thinking (metacognition) has been found to be a useful, self-help tool, particularly when applied to information management, planning and implementing plans, monitoring progress, detecting and addressing misunderstandings, 'debugging' (dealing with errors), and evaluating problem solutions (Haeruddin *et al* 2020). In short, it is to be encouraged.

### 7.3.3 Observe

Another approach is to observe students as they work and note clues to their thinking. This can be time-consuming, it may involve more than one observer, and it may not reveal a lot that is meaningful or useful. It can be useful in judging the individual contributions of students in group work (see below).

## 7.4 Working in groups on practical and other tasks

Assessing the quality of ideas and how they were constructed may seem to lean towards pencil and paper activities and put to one side the assessment of creative practical activity. There is growing interest in breaking away from a diet of 'cookbook experiments' to foster a spirit of enquiry and encourage more creative thinking and activity in laboratory practices (e.g. Dunnett *et al* 2019). The ideal may be for students to bring their own physics problems to solve, explore, construct tentative explanations, and design and carry out practical tests. Given the likelihood that the students' problems will be unexpected and will need unanticipated resources, this approach may not be widely feasible. Perhaps it is more realistic to provide potential problems which draw on available resources and have students work on these problems in small groups.

Experience of collaborative work is not, of course, a bad thing as it often reflects real workplace practices. For instance, a recent and well-publicised challenge to the standard model of particle physics came from a team at the Fermilab near Chicago (Castelvecchi 2021). This team, in excess of 200 members, found that the muon's magnetic moment was larger than the standard model predicted, hinting at the existence of more particles than are currently known (and an explanation of the discrepancy found to be 'far more elegant' than others). But, as far as assessment is concerned, the difficulty is obvious: which group members contributed most usefully and deserve most credit for it? If there is some division of labour (and some division of labour would be usual in large teams), then particular contributions may be credited to individuals. But where minds are all focused on the same aspect of a problem, it is useful to see the task as providing two, overlapping experiences for the student: collaboration and creative problem solving.

### 7.4.1 Collaborative competence

How might we assess the students' ability to work productively in a group? This might entail observation to identify matters of, for instance, group leadership, problem exploration, toleration of others' ideas, and idea development. In each case, there would be an emphasis on the productive functioning of the group (Bowe 2005). For example, Dunnett *et al* (2018) have used 0–3 scores to grade involvement according to students:

- doing all or none of the work (0),
- sharing the work fairly (1),
- actively engaging in discussion (2),
- contributing to effective team work (3).

While these and other criteria may reflect collaborative dispositions and skills to some extent, they do not inform us directly about creative thinking and activity, although they may indicate an environment that supports it.

### 7.4.2 Group creative competence

People can collaborate without thinking creatively. This means that the second experience, collaborative creative thinking, is likely to need attention. Table 7.2 illustrates this for the design of an experiment intended to solve a problem.

The comments specify strengths and weaknesses and this is probably of most use to the students in formative feedback. In practice, it is likely that you would look for several additional, more specific responses. For example: Did the group respond creatively to inadequacies in the design which became evident?

The overall score may be taken as a rough indication of the performance of the group, which, over several problems, may show progress. (Note that the last criterion in the list above is not essential but may add to the score.)

Given that the activity is meant to be a learning experience, it is worthwhile making students aware of what will be valued. They may also benefit from some

**Table 7.2.** Some examples of criteria relating to experiment design. The scores (0, 1, and 2) are intended to reflect the quality of the response: not at all (0); partially sufficient and with potential (1); fully or with only minor deficiencies (2).

| Criterion | Score | | |
|---|---|---|---|
| Did the group construct a design using their own ideas? Comment | 0 | 1 | 2 |
| Was the design appropriate in that it would test the prediction? Comment | 0 | 1 | 2 |
| Was the design impressive (in the sense of elegance, ingenuity, economy)? Comment | 0 | 1 | 2 |

guidance on how collaborative activity can be supported, so it could be useful to have a strategy for doing that. (For example, De Bono's thinking hats approach (outlined below), a strategy found to be effective in industry and in education, can do that.)

This approach may give information on the performance of the group, but as far as the individual is concerned the contribution of particular students may go unnoticed. Did this student contribute something to problem exploration or reformulation, and make plausible suggestions? Were those suggestions mundane or unusual? There is a risk that group work will be shaped by the conscientious few and by strong personalities which suppress idea generation in the others. Observational assessment aimed at catching individual contributions can be very labour intensive if scores for individual students are to be produced for personalised feedback. Nevertheless, students' assessment of their own performance in the group may be useful. If it is a high-stakes activity, however, it could distort student self-evaluations. While a problem solving project offers useful experience of collaboration in physics, and an ability to work in this way could be an asset, it can be difficult to assess individuals other than in terms of their contribution to the group's success.

## 7.5 Risk taking and some caveats

Possibility thinking involves a degree of risk taking. Ideas that seem quite plausible to the student may be judged to be irrelevant by the tutor. Scepticism may also increase with the novelty of the idea—going a long way out on a limb is risky. Balchin (2006) has pointed to the possibility that assessing students' creative thinking will make them play safe. While it may be possible to do some assessment 'by stealth' where students are unaware that they are being assessed (Shute and Rahimi 2021), it seems better and more ethical to be open about it and make clear where the rewards are. Students tune their learning to the requirements of examinations so if they are to value and cultivate their creative thinking, they need to know that it will be judged and the nature of the criteria. Students today come to our universities from all parts of the world, and they may be from cultures which do not value the creative thinking of students in the same way. For instance, putting forward ideas in the face of experts could seem to show a lack of respect for their expertise and knowledge. Equally, this kind of thinking may be taken to show an unwelcome kind of independence. Knowing what possibility thinking is about and why it can be worthwhile is important if such students are to understand why it is expected. Balchin points to other advantages. Creativity in general has attracted popular myths and misconceptions. Bending the mind to matters of assessment helps 'to demystify and operationalise' creative competence. In turn, this can help physicists communicate effectively with one another about this aspect of science, and to evaluate it in what they teach.

There is another issue which can impede the fostering of creative thinking in physics. In the West, there tends to be a culture of negativism—people are critical of ideas even as they are born. The all-too-ready critic and pessimist can suppress creative thinking as it starts. We need the looser, wide-ranging kind of thinking first or there will be no ideas to develop and evaluate (Newton 2016). This means that, at times, thinking in groups has to be managed. Thinking can be guided by De Bono's (1999) six metaphorical thinking hats strategy. Briefly, these hats represent different kinds of thinking and comprise:

- A white hat when information and data collection is appropriate;
- A green hat when creative, constructive thinking is needed;
- A black hat for evaluative, cautious, critical thinking;
- A blue hat when an overview or progress monitoring is needed;
- A red hat for emotional responses to progress;
- A yellow hat when there is the need to identify positives and strengths in the progress.

The leader of a group might choose to begin with a blue hat image and draw out the nature of the problem and the goals, followed by a sharing of white hat tasks. After a review of the information collected, a green hat approach to explore possibilities could be appropriate. Ideas would then be evaluated (black hat). When necessary, participants are reminded that they should not drift from the kind of thinking appropriate to a particular part of the task. In practice, this approach is likely to be somewhat iterative, with some to and fro movement between kinds of thinking, but it has been found to be a useful way of managing engagement with a problem. Wearing one hat, ideas are generated freely; critical thinking is barred until ideas have been collected. Only then is a second thinking hat put on, and the ideas are evaluated for their potential. Synthetic, wide-ranging creative thought and analytical, focused critical thought are separated so that these very different kinds of thinking do not come into conflict. Physics courses with their well-defined problems for students to solve probably do a lot to develop analytical thinking but may give less explicit attention to synthetical thought that brings ideas together to produce something new, at least to the student.

It would be easy to see a lack of creative competence as entirely a personal matter—it's the student's fault. First, the nature of a student's education may have excluded opportunities for creative thinking, and even suppressed them in an emphasis on transmitting a fixed body of knowledge. Surrounded by others with a similar experience, such an outlook is reinforced. As a consequence, the student has lost any tendency to think creatively, or sees it as irrelevant. Second, the environment in which the student works may not favour, value or support creative activity. Creative thinking benefits from opportunity, some autonomy, time, and resources, and the encouragement and example of those who teach. It is impeded by, for example, threatening, harsh evaluation, competition, and excessive workloads (Amabile *et al* 1996, Amabile 1997). Creative behaviours are, in part, down to the student, but they also reflect the ethos of the institution and the attitudes of the tutors (Elton 2006).

## 7.6 Providing feedback

Formative feedback provides guidance on how the student might improve or develop a competence. Ideally, this would combine evidence from the product and process. It might, for instance, direct attention to problem analysis, the use of a particular heuristic or tool, such as a temporal sequence of diagrams, or to the weak plausibility of a potential explanation, or to an inadequate control of variables in an experiment design. Such matters would need attention in subsequent opportunities. The assessment, however, should be seen as an *estimate* of competence. The reliability of assessment may increase over time with a succession of tasks (Treffinger *et al* 2002). As evidence accumulates, it could contribute to a summary of the competences students have shown. They can only show, however, what the tasks allow them to do. If the range is too narrow, then so will be the judgement of competence.

## 7.7 Recognition and measurement

Creative competence in a particular discipline may be more easily recognised than measured. The approaches described above illustrate this but by being clear what you want your students to achieve, and then constructing a test of it, you should be able to collect evidence which will help you provide feedback to your students. In this, it is worth trying to go beyond the product and look at the process behind it. It is the process that students will need to practise and improve if they are to develop in competence. Assessment of that competence is not, however, a clear-cut activity. There are rigorous ways of coding and assessing creative thinking described in research, but they can be time-consuming, labour intensive, and, if applied in general teaching, may discourage rather than promote support for creative thinking. Their relevance to assessing creativity in a discipline such as physics can also be questioned (Sternberg 2018). This is an interesting problem which offers opportunities for creative thinking on the part of the creative physics tutor, and the creative physics tutor is the subject of the next chapter.

## 7.8 Something to reflect on

- In a unit of physics that you teach, identify an ill-defined problem or task which calls for creative thinking by your students. How would you present this problem or task? What would you value in the students' thinking? How would assess what you value?
- Your students are to work as a collaborative, creative group on an assessed task. The task is to produce a tentative explanation of a given event (e.g. a posed problem) and test it by experiment. Some students object to this because they believe their grade will be adversely affected by, for example, the presence of freeloaders. How will your method of assessment reassure them?

## References

Amabile T M 1996 *Creativity in Context* (Boulder, CO: Westview)
Amabile T M 1997 Entrepreneurial creativity through motivational synergy *J. Creat. Behav.* **31** 18–26
Amabile T M, Conti R, Coon H, Lazenby J and Herron M 1996 Assessing the work environment for creativity *Acad. Manage. J.* **39** 1154–84
Balchin T 2006 Assessing creativity through consensual agreement *High. Educ.* **570** 173–82
Bowe B 2005 Assessing problem-based learning *Handbook of Enquiry and Problem Based Learning* ed I Barrett, M Labhrainn and H Fallon (Galway: CELT), pp 103–11
Castelvecchi D 2021 Is the standard model broken? Physicists cheer major muon result *Nature* **592** 333–4
De Bono E 1999 *Six Thinking Hats* (New York: Back Bay Books)
Diakidoy I-A and Constantinos C P 2001 Creativity in physics *Creat. Res. J.* **13** 401–10
Dunnett K, Gorman M N and Bartlett P A 2019 Assessing first-year undergraduate physics students' laboratory practices *Eur. J. Phys* **40** 015702
Elton L 2006 Assessing creativity in an unhelpful climate *Art Des. Commun. High. Educ.* **5** 119–30
Haeruddin H, Prasetyo Z K and Supahar S 2020 The development of a metacognitive instrument for college students to solve physics problems *Int. J. Instr.* **13** 767–82
Kohl P B, Kuo H V, Kowalski S and Kowalski F 2011 Promoting and assessing creativity and innovation in physics undergraduates *Phys. Educ. Res. Conf.* **1413** 39–42
Newton D P 2016 *In Two Minds* (Ulm: ICIE)
Shute V J and Rahimi S 2021 Stealth assessment of creativity in a physics video game *Comput. Hum. Behav.* **116** 106647
Sternberg R J 2018 What's wrong with creativity testing? *J. Creat. Behav.* **54** 20–36
Treffinger D J, Young G C, Edwin C S and Shepardson C 2002 *Assessing Creativity: A Guide for Educators* (Storrs, CT: University of Connecticut)

# Chapter 8

## The creative tutor

### 8.1 The value of creative teaching

So far the emphasis has been on providing opportunities for students to exercise creative thinking and develop creative competences in various ways. We turn here to the creative *tutor*. Clearly, there is a distinction. Creative tutors may be creative in order to achieve a wide variety of goals. One of these may be to provide opportunities for their students to be creative, and we hope it will be. However, creative teaching can serve many ends. Students may have shown that they often have difficulty with, for instance, a particular concept, a mathematical representation, or the application of some instrument in the laboratory. The creative tutor constructs an approach which overcomes the difficulties. From time to time what we are expected to achieve in our discipline may also change. There may be no ready-made solution which fits your particular context so it must come from you and your own creative abilities. This is not at all a bad thing. Teaching in the same way for some 40 years is the road to boredom and discontent for both you and your students. Doing some things in a different way each year can maintain your own interest and add to your job satisfaction. Creative thinking is as much for the teacher as for the student.

### 8.2 Ten questions and answers

Rachel Simpson is a lecturer in education who specialises in the skills of creative teaching and how they can be developed. She kindly agreed to respond to some questions about what creative teaching means in higher education and how it might be developed. This is what she had to say[1]:

---

[1] We thank Rachel for her contribution to this topic and gratefully acknowledge her kind permission to use her responses in this way.

### 8.2.1 What is creative teaching in the context of higher education (HE)?

I think of creative teaching being teaching that has some aspect of novelty for the teacher. This need for novelty is fulfilling a purpose—i.e. creative teaching results in something that is of value. 'What is valued' has to therefore be considered in the context of HE—and then focuses on the purpose of HE. If the purpose of HE is for students to develop their critical and creative thinking skills, for example, then creative teaching will need to enable this. Other graduate attributes can be considered too—e.g. the skill of articulating ideas, collaboration, negotiation, and compromise with peers and future colleagues. In HE these should be underpinned by a sound understanding of the knowledge in the field being studied, to develop informed, reasoned arguments. Again, creative teaching in HE will be pedagogical ways to enable these skills to develop. This might be through two approaches. First, a HE lecturer may use creative thinking skills to plan the content and structure of sessions (e.g. the content of lectures, questions in seminar discussions, activities in workshops). This should align with the context of the student group and also build upon strategies that the lecturer feels comfortable to use and possibly have worked before, in a different context. Strategies that failed are unlikely to be used again. Second, the HE tutor may use creative teaching skills to respond in sessions to the 'unexpected'—students' questions, the multiple perspectives presented, making connections between different ideas generated in tasks. I saw an example of this yesterday when I observed a colleague teach. The theme of the session was 'Identity'. Students responded to the HE lecturer's questions about identity—he asked them questions directly relevant to their lives e.g. 'Which groups do you identify with, and why?' The lecturer didn't know what the students would say, but he listened carefully to each response and skilfully led the discussion by making a comment, connecting different responses, and asking follow-up questions of a new student linked to the previous response. It was evident that the students were being challenged to think deeply and articulate their thinking in the session. The preplanned questions helped with this.

### 8.2.2 What are the benefits of creative teaching?

A main benefit of creative teaching should be for the lecturer's satisfaction and enjoyment—in the professional context. Use of creative teaching skills is an investment, as you will be putting yourself (perhaps your 'professional personality') into the session. The buzz of excitement and interest you have in the session can stay with you for a long time afterwards. As there is an element of openness and risk to creative teaching (i.e. planning open questions, activities, and input that allow the students to engage at a deep level of interaction), it can also be highly stimulating for the lecturer. It can allow new perspectives to be considered and thoughts to develop for the lecturer, and this can be used beyond the session. I see creative teaching as very much a social act—the students involved are a key consideration. Returning to the idea of creative teaching being of value, as well as to lecturer, this value includes gains for the students (graduate attributes, as discussed in question 8.2.1).

Potentially, the lecturer is also modelling creative thinking behaviours for the students they work with—perhaps another gain.

**8.2.3 Is there a downside to teaching creatively?**

From a practical angle, planning intentionally creative sessions can be much more time-consuming and emotionally draining to deliver than sessions that would be seen as more 'traditional delivery'. For example, in HE, if the lecturer is timetabled to deliver a one-hour lecture in a lecture room (fixed seats in rows) to 200 students, a transmissive, information-giving delivery style may be seen as the easiest way forward, with little lecturer–student interaction. This may be because considering a more creative approach could present challenges due to the group size (e.g. controlling the group if discussion tasks are included, the imbalance between numbers—one lecturer versus 200 students can be quite intimidating!) and the practical nature of the session (e.g. room layout). Also, the benefits described above can each have a downside. For example, risk-taking and openness can be highly stimulating but these can also be very challenging in a negative way (i.e. students may disagree with the lecturer, or with each other). Situations of conflict need very skilful management by the lecturer. Such management relies on strong subject knowledge, pedagogical knowledge, and interpersonal skills. Cans of worms can be opened and sometimes a lecturer could feel it is easier to follow a more transmissive lecture style to keep cans of worms closed!

**8.2.4 What would you say are the attributes of a creative teacher in HE?**

Several attributes came to mind for this question. At first I was trying to rank these (is one the most crucial?) but I decided that all are needed, almost in equal measures. So, although the discussion below is not in an order, this was the order in which the attributes came into my mind! I think own curiosity is key—this leads to creative teaching approaches which allow openness to ideas, multiple perspectives, and unpredictable responses, which will hopefully enrich the understandings that develop during a session. Also, creative teachers are unlikely to develop ideas in isolation—curiosity about colleagues' approaches will help them to gain new thoughts. Courage is also key. Teachers in HE may be working in education systems that at least appear to have a traditional (transmissive) approach to education. Although this is changing, there are deep roots of historical practice that a creative teacher needs to contend with. I think that strong collaborative skills can help with courage (working with colleagues to develop creative teaching approaches collectively, or at least gaining others' support). Making connections (perhaps critical/creative thinking skills) is also a key attribute for the creative tutor: if the aim of HE is for the students to develop a deep level of understanding, ability to apply knowledge, and the skills of questioning, reasoning etc, tutors needs to understand contexts (i.e. previous life/education experiences) to help them to engage with their students. Underpinning all of the above would be strong subject knowledge and pedagogical knowledge. There will be many more attributes but a final one that comes to mind is a willingness to make mistakes and be criticised (I'll discuss this later)—this final one perhaps is the biggest challenge!

### 8.2.5 Are these attributes something you are born with, or can they be acquired or developed?

I think these attributes can all be developed and acquired, and this is likely to happen through experience. This may be through:

- First-hand experience of trying something out and it is successful.
- Observing/working with others to gain new ideas.
- Peer support plays a crucial role—if peers praise a session they have observed you teach (for example, using a creative teaching approach) you are much more likely to continue to develop that approach and, in turn, strengthen the attributes connected to the approach.
- I have found that reading about an idea, discussing it with others, and then reflecting on it at a deep personal and professional level can also help to develop an attribute (e.g. developing my own decision-making competences).
- The above focuses on developing creative teaching attributes in a professional context. I do think that childhood experiences play an important role in developing creative thinking skills, which perhaps stay with you beyond childhood. Examples that come to mind for children are opportunities:
    - To take risks (real child-led risks, not those controlled by parents—like climbing trees when there isn't an adult around).
    - To solve real problems.
    - To collaborate with peers and negotiate.
    - To fail at something—I think this is perhaps one of the most important.

There will be other examples but these are the ones that came to mind. I think the creative thinking skills that develop then become deep rooted in you and can be part of your professional identity as adults and tutors.

### 8.2.6 What are the impediments to creative teaching?

I think I have mentioned some already in question 8.2.3. Here are some further thoughts. A main impediment could be the system in which the tutor is working. The tutor wants or needs to be successful—if the education system does not see the value of creative teaching, then it will be a significant challenge. The resistance could come from the immediate education system of the senior leadership team, which is likely to be influenced by a network of factors (such as, assessment systems, perceptions of student satisfaction, student course evaluations, university league tables). Another barrier could be practical—the time it can take to develop a teaching session using a creative approach, rather than simply using last year's lecture slides.

### 8.2.7 Students sometimes see themselves as buying a product rather an education. Do you think this adversely affects a desire to teach creatively? If so, in what way(s)?

I think this links to question 8.2.6 and is potentially a really influential aspect regarding creative teaching. I will use an anecdote to start my answer:

When I came into higher education, I was really looking forward to developing deep levels of thinking with passionate and highly engaged students. I planned my first teaching session carefully, and the session seemed to go well with lots of student interaction and engagement in the topic being discussed. Or so I thought…After the session, I received an email from a student, titled 'A Question About Today's Session'. Before I opened it, I felt excited, and a little nervous. This was my first, student email—I wondered which part of the topic we had discussed they were asking about. I opened the email and read it. It didn't take long, it was one sentence:

'Please can you tell me what I need to do in your module to get a First in the assignment.'

The anecdote above brought my perception of the student experience in HE quickly down to Earth! I was imagining an experience of deep learning but for some (perhaps many) students, the final assessment grade is their main focus. I have only worked in HE during the time when students have paid large tuition fees—some have said directly to me that they are 'paying a lot of money for this'. Hearing this as a university lecturer is hard—I think it creates a feeling of uneasiness—after all, not every student is going to be awarded a First! I think it impacts on motivation to use a creative teaching approach, perhaps in a way that a creative lecturer will need to be 'doubly' creative to find creative ways to balance the two aspects—the students' expectations of a HE experience (the product), and fulfilling the intentions of a HE programme (the processes—which develop the graduate attributes mentioned in question 8.2.1). The HE tutor's role is partly to help students understand the purpose of HE (i.e. graduate attributes). Once this has been established, then the students should see the value of creative teaching approaches. However, I think there is another aspect to consider for this question. If a HE lecturer's creative approach is to encourage openness and risk-taking for their students—this needs to be taken into account in assessments. A student shouldn't be penalised for this so the assessment rubric needs to have flexibility to allow this. This can be challenging as it may not be up to the individual lecturer, as assessment policies are often university-wide. A final thought linked to the quote above—as well as perhaps there being a link to students feeling they are buying an education, this attitude might also link to their previous experiences in education systems. Are GCSE and A-level students told 'what they need to do' to get a top grade?

### 8.2.8 What advice would you give to a new university lecturer about becoming a more creative teacher?

Talk to colleagues; observe colleagues; team teach (to gain multiple perspectives). Take time to have in-depth discussions with mentors—not just about probation targets, but about effective teaching and learning in HE, and what creative teaching really means. Ask for key papers to read—and discuss these with the mentor. Take time to find out about the students—gain understandings of their contexts, needs, future ambitions, and desires.

### 8.2.9 Would you give different advice to someone who is a mid-career lecturer?

This is a really good question! I think the above applies to all but it is likely that mid-career lecturers will be more fixed in the way they approach teaching and learning. It is possible that a change in mindset is needed—this will really depend on the individual's motivation (i.e. why do they need to make this investment?) From a practical view, I would encourage a mid-career lecturer particularly to do the first point in my answer to question 8.2.8—gaining multiple perspectives could really help.

### 8.2.10 Do you see higher managerial colleagues in a university as having a role in fostering creative teaching? If so, what would that role be?

The university's strategy or vision is likely to include graduate aims and attributes that assume creative teaching approaches. This will have been constructed by managerial colleagues. However, how are these graduate attributes monitored and assessed? For example, are collaborative skills (i.e. group work with peers) assessed summatively throughout a degree programme? What about problem-solving skills? Managers could think creatively themselves to consider how developing certain graduate attributes can be seen as something highly valued—this in turn would filter down to lecturers' actions when planning and facilitating effective teaching and learning experiences.

## 8.3 Why teach creatively?

### 8.3.1 Some benefits for the student

Students' strengths, weaknesses, and interests are different, and their backgrounds, experiences, personalities, and traits vary, so what suits one student may not suit another—one size does not fit all, and it probably never did. Moreover, times change, and attitudes, opportunities, and students' goals change with them. Students may no longer see relevance in what is being taught, or else fail to respond as we hoped to how it is taught. Even teaching that was a brilliant success with one group of students can flop with another. It is difficult to enlist a willing engagement with learning when teaching fails to touch what matters to the students. A creative teacher is sensitive to and responds to these differences. This pays dividends, not just in the quality of engagement with learning, but in students' attainment and attitudes (e.g. Gibson 2010).

### 8.3.2 Some benefits for the tutor

Tutors may be taught their teaching skills, or acquire them through some form of apprenticeship, or develop them through trial and error. Creative teachers are able to change and make changes which make their teaching more relevant and effective, and, importantly, maintain their enthusiasm and interest in teaching (Craft *et al* 2014).

At the same time, topics change to reflect developments in a subject. Creative teachers have a competence which can future-proof them because they are able to adapt more readily to change. Moreover, it increases job satisfaction.

## 8.4 Creative teaching to support students' learning

Problems noticed when teaching are opportunities for the tutor to respond creatively and increase students' understanding and skills. Like all problems, they come in a variety of sizes and degrees of complexity. Solutions similarly range from your impromptu responses to students' evident incomprehension, to the construction of whole sessions, or even to the design of a full programme.

For example, as you teach, you see an immediate need for an explanation which is meaningful to your students, so you offer an analogy, a mind-map, or a flow chart. For instance, Einstein's thought experiments and Schrödinger's cat made their messages more meaningful. Newton's reference to a falling apple probably served the same purpose. Analogies, however, have their limits (the Solar System analogy for the atom is an example), but they are useful steps on the way to something better. On a slightly larger scale, a game-like activity may be used to make a point. For instance, the chance nature of radioactive decay may be illustrated using a bowl of dice, each with a one in six chance of coming up with a six, or the use of coloured plastic bricks to illustrate the quark composition of particles (Still 2017). Such strategies can make a session work because they make your communication more effective and increase students' purposeful engagement. At the next level, and usually the result of forethought, you may plan to use some form of flipped teaching, blended learning, a software package, or problem-solving activity, or perhaps a virtual or digital 'escape room' which incorporates some aspect of the physics you teach. Higher still is the overarching level of programme design, perhaps shaped by beliefs about the purpose of a physics education, contemporary practices in physics, and what approaches produce worthwhile thinking dispositions and habits in students. The students can also have parts to play, as when you have them create an analogy or a thought experiment, or find a problem to solve and make into a project, possibly supported by simulations or virtual laboratory software (see chapter 9), and ending with a poster presentation of their conclusions (e.g. Gunawan *et al* 2017).

The process begins, of course, with noting the problem or opportunity. Unless this is simple, it may need clarification: What exactly is the problem? What is the goal? There follows a process of exploration and construction to produce a potential solution, a way of doing things differently. This then needs to be tested. We do not suggest, however, that everything has to be the product of your own mind, and that other people's ideas should be ignored. Creative teaching is not always about re-inventing the wheel, but about noticing a problem and solving it in the best way you can. Bringing ideas from elsewhere to your context, and adapting them to suit it can be a time-saving way of working. Even when they don't fit your situation, they may generate an idea worth trying.

One creative approach to teaching, increasingly popular around the world, with its origins in medical education, is problem-based learning (for a broad background, see e.g. Tan 2003). This is when a problem is presented to a student who then must explore it, draw on a variety of resources to learn more about it, and solve it. An aim of this kind of activity is to develop self-directed learning skills, allow students to work at their own pace and in their own way and, of course, solve problems. (It need hardly be said that these problems are not the kind of mechanical exercises commonly given to students in examinations; instead, they are generally fairly ill-defined and more open.) Usefully, this kind of activity has been found to enhance creative thinking in physics, including generating questions about a topic, possible causes of events, and novel applications of knowledge. As a bonus, it can also improve critical or evaluative thinking (e.g. Sulaiman 2013).

Other people's ideas and teaching aids may have been constructed for a different context and may not quite fit your students' or your needs. Nevertheless, you will usually be able to relate them to your context, and creatively adjust or tune them to suit your resources and situation. When you do construct a teaching idea that is new, like all creative acts, it does not come with a guarantee. Students are diverse and one approach, even a new one, may not suit everyone.

## 8.5 Creative uses of technology

Digital technology, forever evolving, brings new opportunities for the creative tutor to consider (some of which are described in other chapters). Imaginatively integrated with teaching, it may enhance and improve learning outcomes (as in the study using PhET, described in chapter 9). We can distinguish between using technology *to solve teaching and learning problems*, and using technology *to carry some of the burden of thinking* in physics. In practice tutors may wish to combine both. With imagination such tools can often be used in a variety of ways by the individual student or by a group of students working collaboratively.

### 8.5.1 Solving teaching and learning problems

Technology may enhance the quality of learning, as when virtual and augmented reality enable students to 'manipulate' objects and see the effect of their actions on them. In studying electromagnetism, for example, augmented reality makes it possible to overlay field patterns on objects and 'see' the effect of continuously changing variables on the fields and the objects through headsets—in effect, making the invisible visible (Radu and Scheider 2019). Not everything is to do with a student's deep, creative thinking, and there will be times when the creative tutor would like to have a means of teaching a topic in a more interesting and novel way. For instance, an app can allow study to be an 'anywhere activity' using a smartphone (e.g. Arista and Kuswanto 2017). The smartphone has been used in this way to teach elementary rotational dynamics using simulations (for general principles, see Jahnke and Liebscher 2020).

### 8.5.2 Technology taking some of the strain

Technology can relieve some of the tedium, ease the process, and, perhaps, make successful creative output more likely, as when it offers surfaces on which ideas can be displayed, developed, re-arranged, and merged (Buisine *et al* 2007). Collaborative idea generation may be supported by eliciting a group's existing knowledge to construct concept maps, brainstorming using interactive tables and screens, and virtual discussion rooms occupied by avatars, some of which allow anonymity to lessen participants' reserve or apprehension (e.g. Shneiderman *et al* 2006). As a brief illustration, TRIZ (Mann 2002) is a tool for application in a specific domain, namely engineering. It began as a cross-referenced, textual resource focusing on engineering problems and evolved into a software tool. It directs attention to existing solutions to similar problems to stimulate ideas or for adaptation. As is to be expected, the most well-known tools are those that are not specific to one domain, and often these are intended to enhance collaborative activity. For example, De Bono's metaphorical thinking hats (see chapter 7) aims to prompt an appropriate kind of thinking when it is needed so that ideas are generated, refined, and developed. Those that allow ideas to be held 'in the margin' while maintaining a working or production area to develop potential solutions have been found useful. The ideas in the margin are not lost or forgotten and can be brought in at any time. Similarly, being able to move from detail to the whole can help, as can a display which enables the visualisation of relationships (Frich *et al* 2019).

## 8.6 The place of critical/evaluative thinking

Ideas have to be at least plausible and that involves evaluative thinking. Evaluative thinking is a better term to use than critical thinking as the latter tends to induce students to look only for what is negative in an idea. Being negative too soon can kill an idea before its potential has been developed. With forethought, a tutor can set up a no-criticism rule in collaborative idea generation until ideas have been collected.

Given that your teaching is likely to be of some consequence for the students, tutors should probably reflect on and evaluate their own, creative teaching ideas. For instance, they might ask themselves:

- What is new about this approach?
- Is the approach appropriate?
- How likely is it to achieve its goals?
- Could I improve it?
- If it doesn't achieve its goals, what is the backup plan?
- If it is successful, how might I share the idea with others?

## 8.7 Change and challenges

What would encourage creative teaching? Certainly, a tutor who sees value in it is likely to be predisposed to be creative and experiment with, 'What if?' teaching ideas. A favourable ethos in the physics department and the university as a whole could help. Elton (2007) has pointed to the deadening effect of an audit attitude towards teaching, learning, and research. When creative teaching is seen as a valued and expected part of a university's work, and when there is satisfaction in it, physics tutors are likely to explore ideas beyond the conventional (e.g. Grube and Piliavin 2000). This is even more likely if they have experience of creative activity and confidence in their own creative competence (Lee and Kemple 2014). It helps if a tutor is open to change in teaching, is flexible in approach, is willing to take the risk that the seemingly clever idea for teaching a topic may not work, and can show spontaneity and disciplined improvisation while teaching (Gibson 2010). On the other hand, such dispositions, beliefs, and attitudes may be undermined by an educational environment that allows little teacher autonomy, where past practices are the only benchmark of success, where tutors are judged by their ability to reproduce the standard session, and where there is no time for imaginative experimentation, or an acceptance that, at times, it may not work.

Being able to teach creatively has probably always been a valuable competence—repeating the same teaching year after year may never have been what is best for students, or their tutors. But now, this kind of flexible, adaptive teaching may be at a premium. There have been attempts to mechanise teaching from time to time, as with the mechanical and electronic programmed learning machines of the past. Even the textbook could be seen as a teaching machine. Now, AI-supported software may do some of the teaching, but will it be as creative in its responses as a human tutor might be. The tutors' imagination can generate something new, appropriate, and rewarding—characteristics of successful creative thinking in teaching. These characteristics are evident in the creative approaches to teaching physics described in chapter 9.

## 8.8 Some things to reflect on

- *At the small but important level:*
  Think of a teaching session in which you had difficulty helping your students grasp some aspect of the topic. What might you do that is different next time?
- *At the intermediate level:*
  Consider a course you teach. How might you make the students more mentally active, collaborative, and creative?
- *At the overarching level:*
  Assume you had a remit to re-organise an entire year's physics programme. What would your course look like? Why would you make it like that?
- *Or, if you want a different kind of challenge:*
  Thinking in physics can be supported by a variety of models (see, e.g. chapter 3). The role of models is considerable but do students grasp their nature, and consciously try to construct models themselves? How might you develop students' understanding of their nature and role?

# References

Arista F S and Kuswanto H 2017 Virtual physics laboratory application based on the android smartphone to improve learning independence and conceptual understanding *Int. J. Instr.* **11** 1–16

Buisine S, Besacier G, Najm M, Aoussat A and Vernier F 2007 Computer-supported creativity: evaluation of a tabletop mind-map application *Engineering Psychology and Cognitive Ergonomics* vol 4562 (Berlin: Springer), pp 22–31

Craft A, Hall E and Costello R 2014 Passion: engine of creative teaching in an English university *Think. Ski. Creat.* **13** 91–105

Elton L 2007 Assessing creativity in an unhelpful climate *Art Des. Commun. High. Educ.* **5** 119–30

Frich J, Biskjaer M M, Vermulen L M, Remy C and Daisgaard P 2019 Strategies in creative professionals' use of digital tools across domains *Proc. of the 2019 ACM Conf. on Creativity and Cognition (San Diego, CA)* 210–21

Gibson R 2010 The 'art' of creative teaching: implications for higher education *Teach. High. Educ.* **15** 607–13

Grube J A and Piliavin J A 2000 Role identity, organizational experiences, and volunteer performance *Pers. Soc. Psychol. Bull.* **26** 1108–19

Gunawan G, Harjono A and Suranti N M Y 2017 The effect of project based learning with virtual media assistance on students' creativity in physics *Cakrawala Pendidik.* **36** 167–79

Jahnke I and Liebscher J 2020 Three types of integrated course design for using mobile technologies to support creativity in higher education *Comput. Educ.* **146** 103782

Lee I R and Kemple K 2014 Preservice teachers' personality traits and engagement in creative activities as predictors of their support for children's creativity *Creat. Res. J.* **26** 82–94

Mann D L 2002 *Hands-On Systematic Innovation For Engineers* (Kortrijk: CREAX Press)

Radu I and Scheider B 2019 What can we learn from augmented reality? *CHI Conf. on Human Factors for Inquiry-based Learning in Physics (Glasgow, 4–9 May 2019)*

Shneiderman B *et al* 2006 Creativity support tools *Int. J. Hum.-Comput. Interact.* **20** 61–77

Still B 2017 *Particle Physics Brick by Brick* (London: Cassell)

Sulaiman F 2013 The effectiveness of PBL online on physics students' creative and critical thinking *Int. J. Educ. Res.* **1** 1–18

Tan O-S 2003 *Problem-based Learning Innovation* (Singapore: Gale)

IOP Publishing

Creative Thinking in University Physics Education

Douglas P Newton, Sam Nolan and Simon Rees

# Chapter 9

## Creative approaches to teaching physics in the twenty-first century

The use of digital technologies in physics teaching has long been established in the laboratory but has been immensely accelerated by the Covid-19 pandemic. This has made digital teaching and learning evolve rapidly and has opened up new and creative ways to teach physics both in the laboratory settings and in the broader university physics classroom. In this chapter we explore a number of these creative applications including:

*Laboratory learning*
- Interactive simulated experiments.
- At home laboratories.

*Classroom learning*
- Simulation based learning.
- The use of virtual and augmented reality in physics teaching.
- Enhancing peer learning in lectures with technology.

### 9.1 Laboratory learning

Experimentation is a key part of any standard physics degree. The skills gained in the laboratory through active exploratory experimentation garner skills which physics graduates can draw on throughout their professional careers, regardless of their chosen role.

A key challenge in laboratory learning is to allow students the time to reflect on what they're observing so they can make connections with the physics concepts they're learning in lectures. However, achieving this time in a laboratory session can be difficult, as laboratory time is often a scarce resource in the university timetable. In the crucial initial years of a degree, students often don't have the time to engage with experiments at a deeper level and end up focussing on the practical 'press button X, read scale Y' as opposed to the deeper physics at play.

Creative solutions to this problem include providing students with simulated experiences of the experiment so they can engage before a lab session or provide students with a take-home kit so they can engage with the experiments over a longer period outside of the laboratory. Both of these are now addressed in turn.

### 9.1.1 Interactive simulated experiments

As often happens, innovation in physics teaching comes from overcoming a challenge. One of the authors and colleagues became aware of a growing diversity in experimental physics laboratory experience amongst their incoming first year undergraduates. On discussion with the students it became obvious that many had had little laboratory experience during their school years. This was borne about in the literature where in their studies Smithers and Robinson (2007) had noted that 41% of schools they asked that taught up to GCSE level had no physics graduates teaching and that the schools that did not offer significant practical experience and often lacked technical support for the subject.

In our first year degree course, these differences in laboratory physics experience are significant and quickly became obvious in practical classes. When we discussed further with students, we were able to separate students broadly into two groups. The first group, who tended to cope well in our laboratory classes, were from schools where practical science was prioritised with regular weekly practical physics lessons. In particular, these students were skilled in using many types of basic scientific equipment (micrometers, oscilloscopes, etc). However, the second group of students, which was much larger than the first, had had little experience in school beyond key examinable practical physics concepts, which were limited in their scope. These students were less able to easily use basic scientific equipment and this led to increased student anxiety during laboratory sessions.

We were keen to find solutions to this challenge and discussed this openly with colleagues across the sector through the Institute of Physics Higher Education Group. At the time colleagues at the Open University (a distance learning university with a long history in the UK) had been using the work of Kirstein (Kirstein and Nordmeier 2007) to develop interactive screen experiments (ISEs).

ISEs are still image based virtualisations of scientific equipment; Hatherly defines them as 'a highly interactive movie of an experiment, filmed as that experiment was being performed' (Hatherly et al 2009).

They have evolved over the last two decades, following the early work of Theyßen et al, Bacon, and Kirstein and Nordmeier (Bacon 2004, Kirstein and Nordmeier 2007, Theyßen et al 2002). In our work, unlike in earlier work, ISEs were used as pre-laboratory tasks to aid in overcoming a difference in practical physics experience between our two identified student groups.

Pre-laboratory tasks themselves are a well-established tool to help students in engaging in the lab classroom. Johnstone and Al-Shuaili (2001) argue that pre-laboratory work should not simply be reading the manual before attending, instead the preparation should be thoughtfully prepared by the tutor(s) and can have many forms. In particular, it should help the student participate actively, that is, they should think about what they do (Johnstone and Al-Shuaili 2001).

We identified eight experiments and/or pieces of equipment for which ISEs could be created as outputs from this project. These were created to familiarise students with the equipment they would be using in their next laboratory session and varied from the basic but necessary (e.g. calipers, micrometers) to the more sophisticated (e.g. spectrometers, oscilloscopes).

We believe that at the time our production process was unique, in that we worked with students employed as partners in the project who developed these ISEs. Working with students as partners is a well-established practice (Healey *et al* 2016), but here the students were themselves veterans of our lab and were able to bring this lived experience to the design. Our role was to manage the students, to work collaboratively with them on the design and to offer advice and guidance on the appropriate pedagogies and physics to explain.

Using university innovation funding we were able to fund the student developers, buy equipment (hardware and software), and disseminate our findings. Using such funding (which is often available from centres for academic development and the like) is a good way to help develop any creative ideas you have which require the production of something significant. The benefits of using this approach are significant, both in terms of the generation of new learning objects, but also for the student developers. One of our former student developers said that participation was 'one of the most rewarding parts of her time' at university, and that what she did greatly impressed job interviewers subsequently.

The ISEs themselves appeared to be successful, for example, students were asked their level of agreement with the statement:

'The interactive screen experiments (ISEs) provided as preparatory tasks for some of the full experiments helped my understanding of the experiment.'

In both cases over 70% of the students agreed (or strongly agreed) with this statement.

In exemplary open text responses, students stated:

'The screen experiments were really useful, I used them before the lab and when writing up my report to remind myself how the experiment worked.'

Teaching staff commented that they felt the use of ISEs allowed students to reflect more deeply on the scientific concepts involved, rather than becoming side tracked by the technicalities of the equipment they were using, which (after using ISEs) they appeared to be more confident with from the outset.

Many universities have taken this approach further. The Open University, for example, has developed experimental physics modules which utilise a range of ISEs and remotely controlled experiments and don't require the students to physically step inside a classroom as all can be done online (Peachey *et al* 2014). At first glance, there may be some amongst you who find this hard to accept, as experimental laboratory physics is such a core part of an undergraduate curriculum, but if we step back for a moment and consider how we engage with our research, often we control experiments remotely and spend increasingly less and less time in the laboratory as large scale experiments such as the Large Hadron Collider or telescopes such as the Square Kilometre Array become the norm.

However, many of us who teach largely face to face are bought into laboratory-based learning and see it as a vital part of the students physics learning journey. How then do we creatively solve a problem such as the impact of the 2020 Covid-19 pandemic, when in the academic year 2020–21 almost all teaching for parts of the year had to take place online, and what did we learn from this? The answer many universities arrived at, which would have been unthinkable to many in the pre-Covid-19 pandemic times, was to try to replicate the learning outcomes from some of the laboratory classes at home, something explored in the next section.

### 9.1.2 At home laboratories

Responding to the constraints imposed by the Covid-19 pandemic made tutors develop and explore creative ways of doing laboratory work around the world. In this next case study we explore with Dr Aidan Hindmarch, leader of the second-year lab in the department of Physics at Durham, how he led the movement of the electronics lab to at home delivery.

*What was the approach?*
Taking our second-year laboratory module, and moving it to at home delivery. Luckily for us, the module focuses on basically data analysis and statistics taught from a laboratory perspective, starting with utilising basic electronics, which is really a form of circuit set-up for simple experiment. And then of course, moving towards putting all these together into a longer, more detailed experiment where they get to put all of these skills to use in different ways.

Moving this to at home delivery was complicated and resource intensive, but viable, thanks to a lot of hard work from colleagues in the department and the availability of relatively cheap electronics which students could collect from the department and experiment with in their bedrooms. For students at distance, who weren't locally based, we were able to post out the kits.

As the module progresses you move from specific task-based learning to designing your own experiment, utilising data collection from an Arduino based system for which you design the data acquisition code. We were able to get a £70 per student fund from the university to support students buying any additional equipment they needed for their projects.

We used Microsoft Teams extensively to support and give advice to students, and in some ways the real time text chat gave students the ability to get support as and when they needed it.

*What issue were you trying to address when you developed this approach?*
So, in 2020/21 the pandemic meant we didn't have space in the building to run our teaching labs in the way that we wanted to. So as a department we made the decision that we would try to get first year students into the labs as much as possible because we thought that it was important to prioritize them, and we wanted to keep our third year labs open as much as possible because they use a lot more specialist equipment, which is

not easy to do off site. So what that meant was, due to social distancing, we needed to make space for these to happen in our building, which meant we had to move our second-year labs course off site, so that the students did the whole thing at home in the end.

*What were the challenges involved?*
In terms of the students, the nice thing about the course itself is that the general ethos isn't that it's about complicated apparatus. It is that it's about data collection and understanding what you're doing from a sort of basic perspective, which really helped when we moved it to distance delivery. I also felt that I had to really explain this to the students so they got what we were doing and wouldn't feel short changed by the at home delivery. I think that was one of the challenges, making sure the students appreciated that they were meeting the same learning outcomes, so that they could appreciate that what is happening in your kitchen or your living room or your bedroom or wherever would have been happening in a lab.

In terms of the logistics, the most challenging thing was figuring out what equipment we needed, sorting it, and getting it to the students. There were a lot of supply chain challenges brought on by the pandemic, which often made sourcing things difficult.

*How did students react?*
I was pleasantly surprised by the module evaluations that we got back this year (2021/22) because they were really positive. At the start of the pandemic (in 2020/21) scores were good, but I think the students appreciated that what we had done was sort of the best effort under difficult circumstances. Last year (in 2021/22) I think one thing they really appreciated was the fact that they were given the freedom towards the end of the module in the project to pull all of this stuff they'd learned together. There is a lot more freedom in the final experiment and, again, a lot of them did seem to appreciate that sort of flexibility and the freedom to do that.

*How did you adapt what you were doing over time?*
In the academic year 2020/21 the module was delivered through at home labs and online support. In the academic year 2021/22 we moved, due to more relaxed social distancing measures, to a more hybrid approach, where students were in the lab every other week (as opposed to pre-pandemic when the lab was weekly) and worked at home every other week.

Another big thing for us was moving to use microcontrollers such as Arudino's which, for a modest sum of money (much less than laboratory electronics), can do the same job, are portable, and can be adapted to do huge amounts of data capture. Their cost means they can be easily used at home, something which wouldn't be possible with traditional kit.

*How important is creativity in physics to you?*
Research in physics is basically creativity. To really come up with something new that other people are going to be interested in and want to follow and develop there has to be creativity there, to come up with that. Coming up with a theory, coming up with a way of modelling it, coming up with a way of measuring it—all of these things are a form of creativity. For me creativity is so key to becoming a physicist, and as

teachers we need to give our students the opportunity to see this. For me, we provide the students content up until this point, but this module is where students become physicists—that said as module lead I might be biased!

In this case study, there is clearly creativity going on both from the academics in developing and delivering at home laboratory teaching, but also in the innovative way they came up with a project-based learning approach to the final project.

Giving students their own budget, to develop a project using the skills they developed which they could run at home means they'll have learned a significant number of skills, both experimentally and generally (e.g. time and project management). This level of freedom also allows students to explore a number of approaches and bring their creative thinking to the fore. Nevertheless, attention still has to be given to matters of health and safety.

Going beyond laboratory learning, we explore next where and how creative digital solutions can generate new thinking in physics education more broadly.

## 9.2 Simulation based learning

The ability to play with a system or a model is central to the discipline of physics and underpins some creative approaches to teaching physics and enhancing learning. What happens if I disrupt a system by changing one of the variables? For example, doubling the strength of gravity or halving the charge on an electron. You can't do these in the real world but they are possible in models or simulations. Playing in this way can often lead to unexpected results and offer a deeper insight into the underlying physics at work.

Tawil and Dahlan (2017) wanted to improve their students' understanding of quantum physics (Tawil and Dahlan 2017). The felt that the students' test scores were low because the subject (for example, the notion of the quantum, particle waves, and Schrödinger's equation) was abstract, mathematically difficult, and counterintuitive. Their idea was to use an interactive computer package, Physics Education Technology (PhET), to simulate phenomena such as quantum tunnelling, bound states, wave interference, Rutherford scattering, and black body radiation (see the PhET website: http://phet.colorado.edu). The simulations are animated, highly interactive, and let students explore the phenomena themselves.

But they had a question: Could this also develop students' creative performance in, for example, divergent thinking? So, for instance, would this simulation activity increase the number of ideas that the students' produced, would it make their thinking more flexible so that they could adapt and respond as the situation changed, and would the students produce more novel ideas?

To answer these questions they constructed activities to accompany the simulations, they divided the students into two groups (one to test the approach and the other to be taught by the earlier, more transmissive method, without the package), and they set test problems. They found that the simulation experience had a very positive effect on the students' creative responses, and their responses to problem tasks were faster and more effective than those of the control group. In addition, the students were positive about the simulation experience and believed that it helped them understand these aspects of quantum physics.

This way of learning produced more creative responses in the students. This is in line with other observations of the effects of simulation, augmented reality, and digital support which allow students to work at their own pace, engage in self-directed exploration, and experiment with learning materials that enable them to construct their own meanings of events and phenomena. This kind of freedom, including the freedom to make mistakes without unwelcome consequences, supports creative competence, and is worth providing when designing activities to foster creative thinking. It should be said, however, that creative thinking in quantum physics does not guarantee that there will be creative thinking elsewhere. Creative competence can benefit from being exercised in a variety of contexts.

## 9.3 The use of virtual and augmented reality in physics teaching

Many new science and engineering students struggle at first to apply their learned discipline knowledge to everyday situations. The ability to correctly visualise the physical processes at work is for many a barrier to this experience and without it students often struggle. For example:

- Forces in mechanical systems are key to the systems' behaviour yet are invisible to the naked eye.
- Electrons flowing in wires give rise to a range of electromagnetic phenomena, yet we can't see them.
- In quantum mechanics the probabilistic wave function by its very nature can't be measured without collapsing.

If we could see these things (and others), would it deepen students' learning?

The issues here are themselves well represented in the literature; Gilbert for instance notes that visualisation is seen as a key skill in science, yet is often neglected in science education (Gilbert *et al* 2007). One way to overcome this challenge is to use mixed realities presented either through a headset or phone to add an additional layer to normal reality where the additionality provided by the visualisation can be presented. As mentioned in chapter 8, the main two forms of virtual realities available are as follows:

- Virtual reality (VR) is a simulated experience that can be similar to or completely different from the real world which can be experienced through a PC screen or other device, or more increasingly through a VR headset such as a MetaQuest or HTC Vine.
- Augmented reality (AR) is an experience where designers enhance parts of users' physical world with computer-generated input, usually overlayed on to the real world when viewed through a device such as a headset or a phone.

The technologies associated with VR and AR have evolved rapidly over the last decade with prices falling rapidly to around £250–£300 for an entry level headset. If this is still prohibitive most smart phones can be adapted into basic headsets through the use of cardboard based holders such as Google Cardboard, making the experience much more scaleable, particularly when the vast majority of students now have such devices.

Although virtual reality can be useful, it is augmented reality which the literature suggests has most recently been more thoroughly explored by physics teachers in higher education. We'll explore two examples (one VR and one AR) below:

- In their excellent work, Radu and Schneider (2019) used a HoloLens based AR solution to enhance students' understanding of the electromagnetism at play in an audio speaker system. They found that students who saw AR representations of the electromagnetism were significantly more effective in developing an understanding of the structures of magnetic fields and the connection between current and magnetic fields. However, there were many individuals who also fared as well in developing their understanding without the need for AR representations. Although only a small-scale study ($n = 112$), this work is interesting and should be explored further to better understand what makes a situation a good candidate for an AR treatment given the resourcing involved.
- Using the MAROON-VR system on a HTC-VIVE headset, Pirker *et al* have explored with a small set of students ($n = 19$) a two-fold study in which students evaluated their engagement, motivation, usability, and learning with the system (Pirker *et al* 2017). Their results suggested that such systems may be useful as an additional adjunct to in class laboratory experiences and that they present realistic laboratory set-ups in an engaging, interesting, and immersive way and help students to focus more on the learning task when in the actual laboratory.

Some of the examples discussed above are pushing what's possible in the use of digital technologies in learning and teaching. What if, however, you are interested in making your own teaching more interactive but would instead prefer to use established techniques? In the next section we'll explore our final example, the well-established practice of using voting systems and peer instructions in lectures.

## 9.4 Enhancing peer learning in lectures with technology

Peer instruction is a method for teaching challenging concepts that students often misunderstand. A typical class using peer instructions begins with the assignment of pre-lesson work in advance of the session. During the class itself they respond to a question (with discrete options and a correct answer) twice, usually using a voting system, these days using a mobile phone based app or something similar, such as Mentimeter, Turning Point or similar. They answer the question first individually and then after trying to persuade a fellow student with a different response. This method was first developed by Eric Mazur, Professor of Physics and Applied Physics at Harvard (Mazur and Somers 1999), but is closely associated with the flipped lecture approach and is particularly well-researched in physical sciences, life sciences, and maths (Knight and Brame 2018). A schematic for how to use a peer instruction approach is given in figure 9.1.

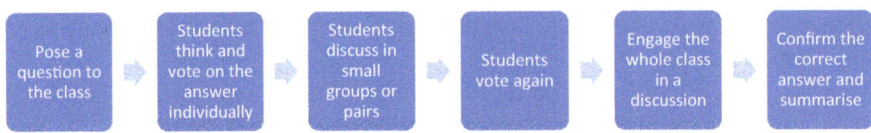

**Figure 9.1.** Schematic showing how to use peer instruction in a class.

It will be a familiar situation to all who teach physics at university that some concepts are challenging to teach because, as an educator, once you have yourself mastered a difficult concept your perspective changes in ways which cannot be unlearned and you may begin to take that concept as self-evident; these difficult concepts are sometimes referred to as threshold concepts in the literature (Cousin 2006). As teachers we may, therefore, find it difficult to explain things we originally internalised many years ago. In contrast our students who have only very recently grasped a concept often teach it well.

Finding the right questions to utilise in a peer instruction approach can be challenging.

An example question (on the topic of classical mechanics for first year undergraduates) is given below:

---

Two wooden balls are the same size, but one has twice the mass of the other. They roll off a horizontal table with the same speed. In this situation:
  a) the heavier ball hits the floor closer to the base of the table than the lighter ball.
  b) the lighter ball hits the floor closer to the base of the table than the heavier ball.
  c) both balls hit the floor at approximately the same horizontal distance from the base of the table.

---

The correct answer here is (c) given that both balls start with zero vertical velocity, they will accelerate under gravity at the same rate, therefore with the same horizontal and vertical velocities, they hit the ground at the same horizontal position from the table.

Pollock *et al* suggest the question must have six key elements (Pollock *et al* 2010):
- *Clarity*—The question is clearly explained so students can understand what is being asked easily.
- *Context*—The question needs to use knowledge which the students have come across before and on a topic covered in class.
- *Learning outcomes*—The question needs to be aligned to the learning outcome of the module and/or session.
- *Distractors*—The 'wrong' answers need to cover common mistakes students make with the topic.
- *Difficulty*—The question needs to be challenging, but viable for the students to solve, e.g. not too easy/hard.
- *Stimulating*—The question needs to engage the students and spark meaningful discussions.

Since Mazur's initial formulation of the method 25 years ago, it has been applied across many disciplines in physics and elsewhere (see Müller *et al* 2017 for a review). What makes the method creative is both the approach itself and the way it makes students act. Those engaging in the approach are more likely to develop new and innovative ways to explain their solution to their peers and as Aidan Hindmarch mentions earlier in this chapter, to 'think like' physicists.

## 9.5 Judging support tools

The quality of what the support tools offer may be assessed on six dimensions (Cherry and Latulipe 2014), namely, in their support for:

- Exploration (how readily the tool enables the exploration of ideas).
- Expressiveness (how much the tool enabled creative ideas to be produced).
- Immersion (how absorbed the user becomes in the subject).
- Enjoyment (how much the user found pleasure in using the tool).
- Results (the extent to which the outcome was worth the effort).
- Collaboration (how much the tool supports collaborative working).

Each of these is assessed on a ten point scale, but asking the students for verbal evaluations may be less tedious and faster. Nevertheless, for all the temptation and promises of such tools and their associated technology, we should not forget Edison's comment about invention: it is 1% inspiration and 99% perspiration. Technology may also be used in quite mundane and conventional ways to reduce some of that perspiration, such as information gathering. Not everything needs deep thought or sophisticated technology.

As a final thought on using AI enabled devices to support teaching, learning, and to augment thinking, these may collect information about the user and use it to adjust their responses. As a matter of ethics, the user should be aware of this, and if data may identify the user and be stored, the user's permission may be needed (Latham and Goltz 2019).

## 9.6 The future

This chapter has illustrated innovative approaches to teaching physics. What they all share is that they all allow the student to investigate something for themselves, be it the virtualised situation described in an interactive experiment, a VR/AR or 2D simulation, or a real-world situation as in an at home laboratory or in a classroom problem such as in peer instruction. Through this process students scope out a new problem space, either physically, virtually or mentally, and gain confidence and knowledge through this exploration, to the point they can feel a sense of justification in their answers to questions about it. This journey from novice to expert is scaffolded but is also an independent and, most critically, a creative journey, with no two students likely to take exactly the same path.

New technologies offer many interesting opportunities for the creative tutor, but we do not want to give the impression that creative teaching is only about finding uses for digital technology. Being enamoured with and an over-dependence on digital technology can risk it coming between the students and the phenomena we want them to understand. Direct experience and the sensing of the world can support meaning and understanding. Nevertheless, this is not always possible and students do have to learn to work with technological aids, so it is a matter of achieving the right balance. To that end, when planning to use digital devices you might consider their advantages and disadvantages. Some things to consider are:

- Have the students some direct experience of the phenomenon (when feasible/possible)?
- What does using the technology add that goes beyond or is better than that of other approaches?
- Does using the technology add usefully to the students' skills?
- How well does a ready-made presentation fit the situation, context, or culture in which it will be used?
- Can the fit be improved?

Thinking of learning physics as a journey also helps us unpick criticisms we've unearthed when speaking with professional physicists. When we've interviewed academics for this book they have all talked about how creative thinking is central to thinking like a physicist. They've also spoken of their own students' journeys and how sometimes the structures and systems in place can hinder this journey. This is best illustrated by theoretical lectures which are still commonplace in many universities and are often detached from work in laboratory classes. When physics curricula get this right the two topics interplay with one another and deepen student's ability to think creatively, but when there is a disconnect between the two students can often be left believing these exist in two separate worlds.

## 9.7 Something to reflect on

The Covid-19 pandemic has led to many changes in practice and, as technology continues to evolve, we'll use it in new and different ways in our teaching.

- Readers might reflect on their own modules and assessments. How do you use technology in your teaching to allow creativity amongst your students?
- In practical terms, how do you seek to evolve your teaching practices to further engage student in creativity activities? How do you reflect on and evaluate what works and what doesn't work in the classroom?

## References

Bacon R 2004 Simulations for physics and astronomy *LTSN Physical Science News* **5**
Cherry E and Latulipe C 2014 Quantifying the creativity support of digital tools through the creativity support index *ACM Trans. Comput.-Hum. Interact.* **21** 1–25
Cousin G 2006 An introduction to threshold concepts *Planet* **17** 4–5
Gilbert J K, Reiner M and Nakhleh M 2007 *Visualization: Theory and Practice in Science Education* vol 3 (Berlin: Springer)

Hatherly P A, Jordan S E and Cayless A 2009 Interactive screen experiments—innovative virtual laboratories for distance learners *Eur. J. Phys.* **30** 751

Healey M, Flint A and Harrington K 2016 Students as partners: reflections on a conceptual model *Teach. Learn. Inq.* **4** 8–20

Johnstone A H and Al-Shuaili A 2001 Learning in the laboratory; some thoughts from the literature *Univ. Chem. Educ.* **5** 42–51

Kirstein J and Nordmeier V 2007 Multimedia representation of experiments in physics *Eur. J. Phys.* **28** S115

Knight J K and Brame C J 2018 Peer instruction *CBE Life Sci. Educ.* **17** fe5

Latham A and Goltz S 2019 A survey of the general public's views on the ethics of using AI in education *Int. Conf. on Artificial Intelligence in Education* (Berlin: Springer), pp 194–206

Mazur E and Somers M D 1999 Peer instruction: a user's manual *Am. J. Phys.* **67** 359–60

Müller M G, Araujo I S, Veit E A and Schell J 2017 A literature review on the implementation of peer instruction interactive teaching method (1991 to 2015) *Rev. Bras. de Ensino de Fis.* **39** e3403

Peachey A, Withnail G and Braithwaite N 2014 Experimentation not simulation: learning about physics in the virtual world *Teaching and Learning in Virtual Worlds* ed C DeCoursey and S Garrett (Oxford: Inter-Disiplinary Press) pp 191–216

Pirker J, Lesjak I and Guetl C 2017 Maroon VR: a room-scale physics laboratory experience 2017 *IEEE 17th Int. Conf. on Advanced Learning Technologies (ICALT)* (Piscataway, NJ: IEEE), pp 482–4

Pollock S J, Chasteen S v, Dubson M and Perkins K K 2010 The use of concept tests and peer instruction in upper-division physics *AIP Conf. Proc.* **1289** 261–4

Radu I and Schneider B 2019 What can we learn from augmented reality (AR)? Benefits and drawbacks of AR for inquiry-based learning of physics *Proc. 2019 CHI Conf. on Human Factors in Computing Systems* pp 1–12

Smithers A and Robinson P 2007 *Physics in Schools and Universities III: Bucking the Trend* (Buckingham: University of Buckingham)

Tawil M and Dahlan A 2017 Developing students' creativity through computer simulation based learning in quantum physics learning *Int. J. Environ. Sci. Educ.* **12** 1830–45

Theyßen H, Schumacher D and von Aufschnaiter S 2002 Development and evaluation of a laboratory course in physics for medical students *Teaching and Learning in the Science Laboratory* (Berlin: Springer) pp 91–104

IOP Publishing

Creative Thinking in University Physics Education

Douglas P Newton, Sam Nolan and Simon Rees

# Chapter 10

## Creating change

### 10.1 Taking the wider view

This book has been about the process of knowledge creation in physics with the intention of bringing its spirit to life in our students and giving it a particular form and direction. Being creative can be very rewarding. Some of the reward is in the delight it generates and the sense of personal effectiveness it confers. Of course, creating knowledge is what the physicist does: the expectation often comes with the job. But a changing world of work can also bring with it a call for a more widespread creative disposition and competence. Having knowledge and understanding continues to be important, but so is this particular kind of thinking which lends itself to problem solving. Physics can offer a remarkable opportunity for its students to experience that, but this will only happen if we provide that opportunity in a coherent and deliberate way. And why hide what we are trying to do from our students? If they know the educational goals and the value of them, they can, at least, see the relevance of what we do and tune their mental activity to it. We do not say that much of current practice is irrelevant—you can't think without something to think about. But we do say that exercising students' creative capacity needs to be brought to the fore. There are many ways to be creative in physics. We have chosen some of them, particularly those which we believe catch the main thrust of the endeavour. We have tried to illustrate how these might be offered to students, but courses vary greatly from one place to another so some (creative) adaptation on the part of the tutor would add to their value.

### 10.2 Some roles

This brings to our attention the role of the physics tutor. Should it be one of transmitting knowledge? Or should it be one of developing thinking competences, creative thinking being one we have marked out for particular attention? Of course,

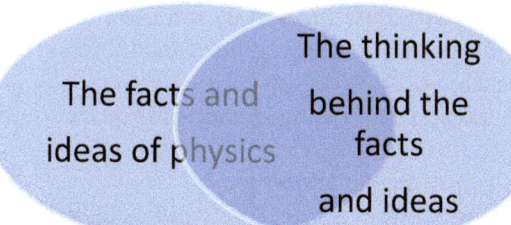

**Figure 10.1.** Knowledge and understanding and thinking competences that produce them, brought together.

it is not one or the other, but a bringing together of both as indicated in the overlap in figure 10.1. There are, of course, other thinking competences we should not neglect, such as analytical thinking, logical reasoning, and evaluative thinking.

When this role focuses on fostering creative thinking, we could attend to:

- *What we might provide*
    - Diverse examples, case studies, or accounts of creative endeavour in physics and making their creative nature explicit (including those that are not paradigm shifting).
    - Tasks that are open-ended (potentially open to different solutions, or ways of reaching the goal).
    - Tasks that, taken together, illustrate the diverse nature of problem solving in physics (rather than only one aspect of it).
    - Tasks that are, as far as possible, authentic (reflecting activity as it is found in physics).
    - Tasks that students tend to find interesting and engaging.
    - Opportunities for both independent and collaborative work.
- *What we might do*
    - Demonstrate creative thinking when teaching by taking part in problem noticing or finding, and problem solving, and describing our own thoughts (novel, appropriate/plausible, satisfying: this 'joint endeavour' may be easier with small groups of students).
    - Welcome ideas and risk taking in creative thinking and demonstrate an acceptance of mistakes which often follow (showing tolerance and respect for the attempts).
    - Encourage students to reflect on their thinking, identify its strengths and weaknesses, strengthen the former and address the latter (particularly in relation to, for example, matters of planning, information management, monitoring progress, misunderstandings, errors, and the quality of solutions).
    - Depict frustration with lack of progress in problem solving as an emotion which tells us a different approach may be needed.

- Progressively withdraw support in order to promote students' independent thinking in order to develop a functioning creative competence in your absence.
- Avoid unnecessarily pressing students for quick solutions, allowing time for ideas to form and develop.
- *How we might adjust the environment*
    - Have students work in a place likely to be conducive to creative thinking (tiered lecture theatres, for instance, can make this difficult, as can places with too much distraction, or those associated with incompatible activities).
    - Make some resources available (for example, large sheets of paper, and/or an interactive whiteboard, or digital tool).
    - If available and relevant, make other kinds of resources available, such as construction kits to model ideas (for example, to foster playful thinking and effective communication).

Each of these can contribute to the fostering of creative thinking (Richardson and Mishra 2018). If colleagues are like-minded about developing this competence in students, it eases course development. Discussion and planning will be needed so that there is a coherent goal, approach, and message. Practical questions to consider are, for instance, how the message is to be communicated to students and how they are to practise it. For example:

- Should there be a short introductory course or 'taster' which introduces students to the meaning and vocabulary of creative thinking in physics, illustrated with case studies and examples?
- Should this be followed by creative activities embedded in the teaching of physics in general?
- How will an overall picture or record of provision be assembled to ensure a broad and coherent provision over time?
- What, if any, are the implications for resources?
- How will students' creative activity be monitored and recognised, and what will this contribute to the overall assessment in the physics programme?
- When will the provision be reviewed for further development?
- How will colleagues be encouraged to value and provide for this aspect of physics?

## 10.3 Some hurdles

### 10.3.1 Inertia and fragmentation

The inertia of past practices, justified by the amount there is to teach, and the lack of time for digression and exploration of questions, puzzles, strange observations, problems, and ideas, can hinder change. If normal practice continues to value and reward *only* analysis and deduction, and 'getting the right answer', then the students will quickly adapt what they do to fit—they see that this is where good examination results will lie. Similarly, opportunities for creative thinking that are uncoordinated and fragmented can lack coherence, impact, and long-term effect. In an ideal world students would begin with noticing some peculiarity of interest and see it through to its solution. In reality, this may not often be possible, but students should at least learn that there is a bigger picture, even if they are only to craft parts of it. They should not complete their degree and be able to say that nothing they did was creative, or that disconnected bits here and there offered only disjointed and incomprehensible opportunities. Talking and planning with colleagues who are open to teaching ideas and presenting successful case studies at teaching conferences helps to develop and disseminate change. This calls for a vocabulary to facilitate the discussion. We have tried to develop one here for talking about the creative endeavour in physics, and to establish a creative learning environment that fosters it in students, but that is not the end of it—people can use the same words but mean different things by them.

### 10.3.2 Notions of creativity

It is important that colleagues agree on what will count as creative thinking. For example, most will agree that if we think creatively, it is with the intention of producing something more or less different or new. But what is 'new' may turn out to be something that is new and clever only to us, and not to others. For a professional physicist, this matters—there is little credit attached to constructing something that already exists. Students, however, are not yet professional physicists and we should recognise creative thinking that is not entirely new to the world or of huge consequence. In addition, there needs to be an appreciation that creative thought does not come with a guarantee. Creative thought can exist without always producing an entirely plausible solution, and sometimes it leads us down a blind alley, but we learn from it. Given that, we may need to ask colleagues if it is appropriate to recognise creative thinking *only* in the successful solution. At the same time we have tried to show that problems in physics can vary widely, from constructing and labelling a concept, to explaining a puzzling event, to testing that explanation, and possibly applying it elsewhere. We see these as worthwhile opportunities for creative activity, but there needs to be some agreement about where the emphasis should or could be in a particular unit or course. Agreement amongst colleagues about such matters in important for course design, but also for assessment. Tutors with different notions of what counts can produce different grades for the same piece of work.

### 10.3.3 Notions of the source of creative abilities

Another aspect which can affect the fostering of creative competence is how people see its source. Some may see it as being wholly innate—something we are or are not born with. Some students say that they are 'just not creative'. When tutors also believe this, they may see little point in making it a learning goal, or even in practising it. Others may take a different view and see it as something which can at least be practised and sharpened in a particular context to tailor any inherent abilities to suit the subject. Still others may see the human mind as being very flexible and capable of developing additional thinking skills. The attitudes these views generate can lead to very different levels of enthusiasm for, commitment to, and expectations of a physics education. Beliefs can be simplistic, misleading, and can misdirect effort. It is not a matter of abilities being due to nature (arising from innate, genetic, and brain physiology) or nurture (arising from experience, learning, and culture). Even if some students have been favoured by genetics or experience, the interaction of nature and nurture is complex and produces students with a variety of strengths and weaknesses, not just those who can and those who can't think creatively. (Sasaki and Kim (2017, p 6) point out that, 'it is more accurate to say that nature and nurture each contribute 100% to the equation'.) Students of physics may have shown some aptitude for the subject, but this is not to say that the aptitude has reached its potential, or that a creative disposition cannot be encouraged, and creative thinking cannot be practised, tuned to context, or otherwise improved. One-sided views of mental aptitudes may need to be discreetly countered if all tutors are to see useful purpose in fostering their students' creative competence.

### 10.3.4 Students' notions of creative physics

There also needs to be some awareness of students' notions of the role of creative thinking in the sciences. On the one hand, they may have none. On the other, they may have acquired, unconsciously or otherwise, a view that the process of generating knowledge in physics is a matter of explaining an event, show the explanation to be true, and applying it in elsewhere. That it involves a fallible imagination, tentative ideas, and a long route to general acceptance may not figure greatly in a student's consciousness. For some, problem solving may be seen in simplistic ways and without a need for imagination—some may even shrink from an expectation that they should think creatively. Colleagues need to be aware that students may arrive with various notions of creative thinking, and of its presence in physics and in learning about physics.

## 10.4 Health, safety, and risk assessment

A consequence of having students think creatively is that it can lead to activities you did not plan or foresee. **As with any other activity that we ask students to do, these need to be assessed for risk, and appropriate measures taken to eliminate the risk or reduce it and its consequences to an extremely low and inconsequential level. Consideration of health and safety matters is important and is your responsibility. Due regard should be given to the level of knowledge and expertise of the student and adherence to your institution's safe working expectations, procedures, safety rules, and practices.** This applies to working with students' ideas, and also to practical ideas which stem from other sources, including those mentioned in this book. If you also have your students engage in these risk assessments themselves, they can become a part of the students' learning experience, and also help to prepare them for practical work in their employment after graduation.

## 10.5 Physics as a dynamic discipline

Physics does not stand still, but continually adds to knowledge and revises, adjusts, or replaces what has gone before. At the same time, ways of thinking and working in physics are not static. Digital technology, for instance, is now important in data handling, modelling and simulation, and provides ways of constructing alternative worlds to test potential solutions to problems. Computational thinking also offers ways of solving problems, such as by iteration, a common strategy in digital processing. Changes like these will continue to add opportunities for tutors to be creative in their teaching. If we don't present physics as an active, creative, developing subject, students may not see it as that.

Furthermore, physics, like some other disciplines, tends to fall into discrete, self-contained packages, and so we teach it in that way. It is convenient and resources have been developed to support that way of working. However, in the eyes of the student physics is not a coherent whole, but a collection of unrelated, disconnected topics, and the transfer of knowledge and connections between them is, at best, limited. It is in such transfers and connection-making that opportunities for creative thinking exist. For instance, there are parallels, analogies, and patterns which are common to both electrical and thermal conduction. These bring them closer and can enhance personal understandings, a first step towards creative production. More than that, the knitting together of disparate parts of physics into a coherent whole (perhaps, to bring quantum physics and astrophysics closer together) could be seen as a worthwhile goal. Nevertheless, we set our students up for a perpetual dissociation between them. We suggest that there should be some imaginative course design which encourages the crossing of boundaries and which changes what, in our students' eyes, is open to their imagination.

## 10.6 Creative physics and the cultivated imagination

Ideas move physics on. They are central to it and have many forms, but reason helps to shape, constrain, and test them, and, in the end, it rejects many. But ideas are about what might be or could be and begin in the imagination. Imagined worlds are central to creative thought. Some may even disrupt the thinking landscape and lead to further ideas. For too long, we have allowed the notion of creativity to be confined to a narrow range of human endeavour, like the arts, as though it did not exist elsewhere. It is not a threat to the rigour or objectivity of science—it is what gives science something to think rigorously and objectively about. Some may claim that they expect students to be creative thinkers, but it is often not an objective of the course, neither is the expectation made explicit, nor the opportunities made clear, and little, if anything, is done to develop students' creative competence directly. It is time to claim the various creative behaviours for physics, to make our students aware of their importance and relevance, and to help them retain and cultivate their scientific imagination.

# References

Richardson C and Mishra P 2018 Learning environments that support student creativity *Think. Ski. Creat.* **27** 45–54

Sasaki J Y and Kim H S 2017 Nature, nurture, and their interplay *J. Cross-Cult. Psychol.* **48** 4–22

www.ingramcontent.com/pod-product-compliance
Ingram Content Group UK Ltd.
Pitfield, Milton Keynes, MK11 3LW, UK
UKHW050244150426
5217IPUK00005B/124